Behavioral Mitigation of Cooking Fires Through Strategies Based on Statistical Analysis

Behavioral Mitigation of Cooking Fires Through Strategies Based on Statistical Analysis

Final Project Report for EME-2005-CA-0343

Marty Ahrens
John Hall
Judy Comoletti
Sharon Gamache
Amy LeBeau
National Fire Protection Association

August 2007

Table of Contents

· ·

List of Tables and Figures

continued on next page

Executive Summary

Cooking Fires and Injuries: The Size of the Problem

Cooking equipment was involved in 31 percent of home structure fires reported in 2003.

Cooking equipment, most often a range or stovetop, is the leading cause of reported home fires and home fire injuries in the United States. Cooking equipment is also the leading cause of unreported fires and associated injuries or illnesses. When cooking equipment is described as a cause, it means that cooking equipment provided the heat that started the fire, not that the equipment malfunctioned. More cooking equipment fires are caused by human error than equipment malfunction.

In 2003, U.S. fire departments responded to 118,700 home structure cooking equipment fires. These fires caused 250, or 8 percent, of the home civilian fire deaths; 3,880, or 29 percent, of reported home civilian fire injuries; and $512 million, or 9 percent, of the associated direct property damage. The vast majority of cooking fires, however, are handled privately and are never reported to the fire department. The majority of reported home cooking fires also were small. From 1999 to 2003, 71 percent of the reported cooking fires were coded as either confined cooking fires or as having flame damage confined to the object of origin. Even so, 38 percent of the reported injuries and 8 percent of the fatalities resulted from these small fires. These statistics are national estimates derived from the U.S. Fire Administration's (USFA's) National Fire Incident Reporting System (NFIRS) and the National Fire Protection Association's (NFPA's) annual fire department experience survey.

Although reported cooking fires and associated injuries and property damage show very similar patterns by time of fire, the pattern for cooking fatalities more closely resembles that of other home fire fatalities. Forty-one percent of the people killed in U.S. home cooking fires from 1999 to 2003 were sleeping when fatally injured.

Findings by Gender and Age

Males face a disproportionate risk of cooking fire injury relative to the amount of cooking they do.

Although women do the majority of the cooking and were the cooks in most of the fires in studies that examined gender, more than half of the people killed and almost half of those injured in reported cooking fires were male. Little gender difference is seen in terms of activity at time of injury. Fifty-six percent of the males and 54 percent of the females injured in cooking fires were hurt while attempting to fight the fire themselves.

Young children and older adults faced a higher risk of death from cooking fires than did other age groups.

Children under five and adults over 65 face a higher risk of death from fires of most causes, including cooking. People 25 to 34 years of age faced the highest risk of cooking fire injury. Youths and young adults 15 to 24 years of age, adults aged 35 to 44, and those 75 years of age or older also faced an elevated risk of cooking fire injuries.

Young children were at high risk from non-fire cooking-related burns.

Although young children are not at high risk for cooking fire injuries, their risks of thermal burns and scalds from cooking equipment, cookware, tableware, or hot foods or beverages are very high. Children may be injured when they reach and pull down on a cord or container, when they run into or are run into by an adult carrying something hot, or when they touch hot cooking equipment, cookware, or tableware.

Findings Related to Leading Factors in Home Cooking Fires and Losses

Unattended cooking is the single leading factor contributing to cooking fires.

From 1999 to 2003, cooking equipment had been left unattended in 37 percent of the home cooking equipment fires reported in Version 5.0 of NFIRS. In addition, unattended equipment was a factor in 42 percent of the cooking fire deaths and 44 percent of the injuries. The share of fires resulting from unattended equipment varied by the type of cooking equipment involved. While unattended equipment was a contributing factor in 37 percent of the reported cooking fires overall, it was a factor in 45 percent of the deep fryer fires and 43 percent of the range fires. It was cited as a factor in only 21 percent of the conventional oven or rotisserie fires and 17 percent of the microwave oven fires.

People who begin cooking when drowsy, impaired by alcohol or drugs, or otherwise limited may be more likely to stop paying attention to that cooking inadvertently.

Properly maintained smoke alarms also provide important protection against fires that occur when the cooking is forgotten or the cook falls asleep.

Many other cooking fires begin because combustibles are too close to cooking heat sources.

Some type of combustible material too close to the cooking equipment was a factor in 13 percent of home cooking fires, 24 percent of the associated deaths, and 12 percent of the associated injuries, making heat source too close to combustibles the second leading factor contributing to ignition for home cooking fires, after unattended equipment. Combustibles include loose clothing, potholders, oven mitts, wooden utensils, paper or plastic bags, food packaging, towels, or curtains. Clothing is rarely cited as the first item ignited in a cooking fire, but it accounts for 8 percent of total home range fire civilian deaths, a comparatively high share. It also has a much higher rate of both fatal and non-fatal injury per 100 fires than other cooking fires. Certain types of clothing, including garments with loose, flowing or dangling sleeves, present an elevated risk of contact with and ignition by cooking heat sources. Older adults were at higher risk of both fatal and nonfatal injury from this type of incident than people of other ages.

Frying is the cooking method posing the highest risk.

Because unattended cooking is cited less often and may have less severe consequences for some types of cooking equipment compared to others, it may be useful to address unattended cooking in part by steering cooks—especially those whose conditions make unattended cooking more likely—toward types of cooking that are more tolerant of unattended cooking.

Frying accounts for a majority share of cooking fires in the few studies that identify cooking method. Frying fires typically occur early in the cooking activity and, if fire occurs, the cooking equipment is typically open and will not contain the fire. Finally, frying employs a combustible medium—cooking oil or grease—which is the first item ignited in most frying fires. No other cooking method has a risk comparable to the risk of hot oil. Hot oil also poses a scald burn risk. For all these reasons, there can be no exceptions to attendance at frying by the cook.

Broiling and grilling do not inherently involve a combustible medium in addition to the food. However, both types of cooking often involve a need for regular cook intervention, such as turning the food in order to avoid overheating. As a result, both methods of cooking can be regarded as only slightly less risky than frying.

Baking and roasting do not inherently involve a combustible medium in addition to the food and typically are done in an oven, which provides containment for fire if one begins. Primarily for this last reason, baking and roasting can be regarded as less risky than broiling and grilling. Brief absences during baking and roasting, which tend to take longer than frying, broiling, or grilling, can be justified, provided a timer is used to remind the cook to check on the cooking.

Toaster ovens can be regarded as small baking devices, although they can be used for broiling as well. Hot plates and food warmers involve conducted heat rather than convective heat. Together with toasters and toaster ovens, they account for most of the fires and related deaths associated with portable cooking or warming devices. Hot plates and toasters should not be left unattended during their typically very short cooking periods.

Boiling does not inherently involve a combustible medium in addition to the food. In fact, the normal medium of water will typically prevent fire until or unless it boils away. Boiling does not normally involve a need for regular cook intervention. Primarily because few fires occur early in the boiling process, boiling can be treated as comparable to or less risky than baking and roasting. Brief absences during cooking can be justified, provided

a timer is used to remind the cook to check on the cooking.

Heat levels for slow cooking are typically low enough that other provisions for safety, including close attendance, are not necessary. If the cookware is placed where an unlikely minor overflow will not contact other combustibles, there will be added safety. If a crock pot or similar device is used, any ignition of food also will be contained, provided nothing has interfered with the equipment itself.

More than half of the home cooking injuries occurred when people tried to fight the fire themselves.

Fifty-five percent of the people who were injured in U.S. home cooking fires from 1999 to 2003 were injured when they tried to fight the fire themselves. This is a far higher percentage than is seen from fires of other causes.

For civilians injured while fighting the fire, only a 4 percentage point difference was seen between the 65 percent share who had been in the area when the fire started and the 61 percent who were injured in fires resulting from unattended cooking. In other words, being in the cooking area versus in another room made little difference in the type of injury a person could suffer if he or she were injured.

More than one-third of the reported cooking fire injuries resulted from fires that were either confined to the object of origin or had an incident type indicating a confined cooking fire. These injuries probably cannot be prevented unless the fire itself is prevented.

The evidence suggests that when confronted with a minor fire, many, if not most, will handle it themselves. So while it is safest to get away from the fire and outside of a burning structure, it would be appropriate to devote some educational resources to teaching more people how to fight fires safely and effectively. Guidelines to help assess the danger of the situation may be useful.

However, there are many messages, often contradictory, in circulation about the best way to handle kitchen fires. These messages can leave people unsure about how to proceed or even lead to demonstrably unsafe firefighting practices that will make the situation worse rather than better. Unfortunately, there is little detailed research on the relative effectiveness or the relative injury risks associated with different approaches to handling small fires. As a result, many of the decisions required to develop consistent, sound, and realistic advice on how to handle and possibly fight cooking fires, must be made as the best judgments of experts rather than definitive research directly on point.

The consensus is clear that water should never be used on a grease fire or on fires with electrical components. But while some experts recommend using baking soda or salt on certain fires, others consider this impractical or even dangerous. Smothering a fire with a lid seems to be an accepted approach. And, while the possibility of burns exists, a properly selected pan lid can cover the fire in one motion and can be used to shield the hand and arm of the resident while the lid is being put in place. In addition, fire blankets are routinely recommended in Europe and Australia but less often mentioned in the U.S.

Fire extinguishers also are recommended often, but when used incorrectly, they can actually spread a fire. It is important that individuals who would consider using a fire extinguisher in a fire situation receive training in how to use these devices properly. It is also important to ensure that this equipment is properly maintained and operational. Many of the sources available mention fire extinguishers in passing, but most provide little specific guidance on how to use such equipment. While hands-on training is the best way to learn to use fire extinguishers, it is likely that many people who have these devices have not received any kind of training at all on their use.

Findings on Program Effectiveness

Educational effectiveness may be enhanced by linking burn prevention and fire prevention.

Traditional fire safety education has focused on preventing fires. Scald and contact burns seem to be close relatives to fire burns. In fact, many scald burns from hot oil occur when the oil spills on individuals carrying flaming pans. The most effective way to prevent a scald burn from burning oil is to prevent the oil from igniting. Given that time is scarce for both life safety educators and the public, and that fire prevention and burn prevention messages are similar and likely to be geared to the same audience, it makes sense to combine these efforts when possible. When advising parents to keep young children away from the stove area, it also is logical to advise the parents to keep children out of the traffic patterns when hot food is being transported, and to keep hot dishes and beverages out of children's reach.

It is also possible that a more holistic approach to prevention will help our audiences better understand the potential dangers and extrapolate safety practices from the messages to their own unique circumstances. It can be hard to find the underlying logic associated with a series of brief, independent messages, particularly when related hazards are not addressed.

Technology may be the best long-term solution to dealing with the cooking fire problem.

The fire safety community has been advising people to avoid unattended cooking for decades, yet unattended cooking remains the leading factor contributing to these ignitions. Technological solutions that either shut off or turn down stoves when no motion is detected, or before a burner can get hot enough to start a fire, may offer the opportunity to improve safety without major changes in a behavior that has proven resistant to change for so long.

Cooking Fire and Burn Prevention Behavioral Mitigation Messages

The following educational messages for safe home cooking to avoid fires and other burns have been developed based on the research findings of this project:

Choose the right cooking equipment. Install and use it properly.

* Always use cooking equipment tested and approved by a recognized testing facility.

* Follow manufacturers' instructions and code requirements when installing and operating cooking equipment.

* Plug microwave ovens or other cooking appliances directly into an outlet. Never use an extension cord for a cooking appliance, as it can overload the circuit and cause a fire.

Watch what you heat!

* The leading cause of fires in the kitchen is unattended cooking.

* Stay in the kitchen when you are frying, grilling, or broiling food. If you leave the kitchen for even a short period of time, turn off the stove.

* If you are simmering, baking, roasting, or boiling food, check it regularly, remain in the home while food is cooking, and use a timer to remind you that you're cooking.

Stay alert.

To prevent cooking fires, you have to be alert. You won't be if you are sleepy, have been drinking alcohol, or have taken medicine that makes you drowsy.

Use equipment for intended purposes only.

Cook only with equipment designed and intended for cooking, and heat your home only with equipment designed and intended for heating. There is additional danger of fire, injury, or death if equipment is used for a purpose for which it was not intended.

Keep things that can catch fire and heat sources apart.

* Keep anything that can catch fire—potholders, oven mitts, wooden utensils, paper or plastic bags, boxes, food packaging, towels, or curtains—away from your stovetop.

* Keep the stovetop, burners, and oven clean.

* Keep pets off cooking surfaces and nearby countertops to prevent them from knocking things onto the burner.

* Wear short, close-fitting or tightly rolled sleeves when cooking. Loose clothing can dangle onto stove burners and can catch fire if it comes in contact with a gas flame or electric burner.

Know what to do if your clothes catch fire.

If your clothes catch fire, stop, drop, and roll. Stop immediately, drop to the ground, and cover face with hands. Roll over and over or back and forth to put out the fire. Immediately cool the burn with cool water for 3 to 5 minutes and seek emergency medical treatment.

Know what to do if you have a cooking fire.

* When in doubt, just get out! When you leave, close the door behind you to help contain the fire. Call 9-1-1 or the local emergency number after you leave.

* If you do try to fight the fire, be sure others are already getting out and you have a clear path to the exit.

- Always keep an oven mitt and a lid nearby when you're cooking. If a small grease fire starts in a pan, smother the flames by carefully sliding the lid over the pan (make sure you are wearing the oven mitt). Turn off the burner. Do not move the pan. To keep the fire from restarting, leave the lid on until the pan is completely cool.

- In case of an oven fire, turn off the heat and keep the door closed to prevent flames from burning you or your clothing.

- If you have a fire in your microwave oven, turn it off immediately and keep the door closed. Never open the door until the fire is completely out. Unplug the appliance if you can safely reach the outlet. After a fire, both ovens and microwaves should be checked and/or serviced before being used again.

Prevent and treat scalds and burns.

- To prevent spills due to overturn of appliances containing hot food or liquids, use the back burners when possible and/or turn pot handles away from the stove's edge. All appliance cords need to be kept coiled and away from counter edges.

- Use oven mitts or potholders when moving hot food from ovens, microwave ovens, or stovetops. Never use wet oven mitts or potholders as they can cause scald burns.

- Replace old or worn oven mitts.

- Treat a burn right away, putting it in cool water. Cool the burn for 3 to 5 minutes. If the burn is bigger than your fist or if you have any questions about how to treat it, seek medical attention right away.

Protect children from scalds and burns.

- Young children are at high risk of being burned by hot food and liquids.

- Keep young children away from the cooking area by enforcing a "kid-free zone" of 3 feet (1 meter) around the stove.

- Keep young children at least 3 feet (1 meter) away from any place where hot food or drink is being prepared, placed or carried. Keep hot foods and liquids away from table and counter edges.

- When young children are present, use the stove's back burners whenever possible.

- Never hold a child while cooking, drinking, or carrying hot foods or liquids.

- Teach children that hot things burn.

- When children are old enough, teach them to cook safely. Supervise them closely.

Install and use microwave ovens safely.

- Place or install the microwave oven at a safe height, within easy reach of all users. The face of the person using the microwave oven should always be higher than the front of the microwave oven door. This is to prevent hot food or liquid from spilling onto a user's face or body from above and to prevent the microwave oven itself from falling onto a user.

- Never use aluminum foil or metal objects in a microwave oven. They can cause a fire and damage the oven.

- Heat food only in containers or dishes that are safe for microwave use.

- Open heated food containers slowly away from the face to avoid steam burns. Hot steam escaping from the container or food can cause burns.

- Foods heat unevenly in microwave ovens. Stir and test before eating.

Use barbecue grills safely.

- Position the grill well away from siding, deck railings, and out from under eaves and overhanging branches.

- Place the grill a safe distance from lawn games, play areas, and foot traffic.

- Keep children and pets away from the grill area by declaring a 3-foot "kid-free zone" around the grill.

- Put out several long-handled grilling tools to give the chef plenty of clearance from heat and flames when cooking food.

- Periodically remove grease or fat buildup in trays below grill so it cannot be ignited by a hot grill.

- Use only outdoors! If used indoors, or in any enclosed spaces, such as tents, barbecue grills pose both a fire hazard and the risk of exposing occupants to carbon monoxide.

Charcoal grills

- Purchase the proper starter fluid and store out of reach of children and away from heat sources.

- Never add charcoal starter fluid when coals or kindling have already been ignited, and never use any flammable or combustible liquid other than charcoal starter fluid to get the fire going.

Propane grills

- Check the propane cylinder hose for leaks before using it for the first time each year. A light soap and water solution applied to the hose will reveal escaping propane quickly by releasing bubbles.

- If you determined your grill has a gas leak by smell or the soapy bubble test and there is no flame:

  - Turn off the propane tank and grill.

  - If the leak stops, get the grill serviced by a professional before using again.

  - If the leak does not stop, call the fire department.

- If you smell gas while cooking, immediately get away from the grill and call the fire department. Do not attempt to move the grill.

- All propane cylinders manufactured after April 2002 must have overfill protection devices (OPDs). OPDs shut off the flow of propane before capacity is reached limiting the potential for release of propane gas if the cylinder heats up. OPDs are easily identified by their triangular-shaped hand wheel.

- Use only equipment bearing the mark of an independent test laboratory. Follow the manufacturers' instructions on how to set up the grill and maintain it.

- Never store propane cylinders in buildings or garages. If you store a gas grill inside during the winter, disconnect the cylinder and leave it outside.

Have working smoke alarms.

- Install smoke alarms in every sleeping room, outside each sleeping area, and on every level of your home. For the best protection, interconnect all smoke alarms throughout the home. When one sounds, they all sound.

- Test each smoke alarm at least monthly.

- Install a new battery in all conventional alarms at least once a year.

- If the smoke alarm chirps, install a new battery in a conventional smoke alarm. Replace the smoke alarm if it has a 10-year battery.

- To prevent nuisance alarms, move smoke alarms farther away from kitchens according to manufacturers' instructions and/or install a smoke alarm with a pause button.

- If a smoke alarm sounds during normal cooking, press the pause button if the smoke alarm has one. Open the door or window or fan the area with a towel to get the air moving. Do not disable the smoke alarm or take out the batteries.

- Treat every smoke alarm activation as a likely fire and react quickly and safely to the alarm.

Introduction

· ·

Fires resulting from cooking continue to be the most common type of fire experienced by U.S. households. This is true for fires reported to fire departments and those handled by private individuals. Cooking fires are also the leading cause of home fire injuries. As a result, the U.S. Fire Administration (USFA) has partnered with the National Fire Protection Association (NFPA) "to research the types of behaviors and sequences of events that lead to cooking fires and develop sound recommendations for behavioral mitigation strategies that will reduce such fires and their resultant injuries and fatalities."

This study of the causes of cooking fires and cooking injuries and the effectiveness of strategies to prevent them also considers as part of its scope cooking burns of all types from all types of products involved in preparing and serving food or drink. Although many cooking injuries result from knives or broken glass and many people are made ill by improperly handled food, these other issues are beyond the scope of this project.

An extensive literature review on cooking fires and burns was conducted to provide the broadest possible fact base for recommendations. This review used internet searches on cooking fires and cooking burns, as well as searches of USFA's Learning Resource Center, the U.S. Consumer Product Safety Commission's (CPSC) Web site, and NFPA's Web site to identify information sources. Information also was sought through direct contact about specific programs addressing cooking safety.

In addition, statistical analyses of data collected by USFA's National Fire Incident Reporting System (NFIRS) and NFPA's annual fire department experience survey provided national estimates about the circumstances and victims of cooking fires reported to U.S. fire departments. NFPA's statistical analysis of cooking fires used Version 5.0 NFIRS data only for the analyses from 1999 to 2003.

NFIRS is the most representative national fire database, providing detailed information on individual fires and casualties. Nearly all national estimates of specific aspects of the U.S. fire problem begin with NFIRS. Roughly half to two-thirds of U.S. fire departments—working through their respective States—participate in NFIRS, which currently receives reports on more than one-half of the fires reported to local fire departments each year. The NFPA and most other users of NFIRS combine it with the NFPA survey to produce the best "national estimates" of the specific characteristics of the U.S. fire problem. Any unreferenced fire statistics in this report are national estimates from NFIRS and the NFPA survey produced by NFPA staff. See Appendix A for more details.

Statistical analyses of data collected by CPSC's National Electronic Injury Surveillance System (NEISS) also were conducted. NEISS tracks injuries that were treated in a sample of roughly 100, or 2 percent, of hospital emergency rooms.[1] This information has been used to develop projections of injuries caused by products and to identify unsafe products or practices when using the products. In recent years, its scope has expanded to include all injuries. Brief narrative information is available on incidents in the sample. This information helps to illustrate more fully the mechanism of injury. Unreferenced statistics in this report from

CPSC's NEISS also are based on analyses done by NFPA staff.

Fire department reports on cooking fires collected by NFPA's Fire Incident Data Organization (FIDO) also were reviewed. However, because they provided little new information, the reports from FIDO were not used.

A draft report of the completed literature review and statistical analyses was provided to NFPA's Educational Messaging Advisory Committee (EMAC), an ongoing group of volunteers that exists independent of this project, to review and revise cooking fire educational messages, based on the research. Because of the large number of issues and related findings, EMAC concentrated on what were considered the most important issues, which included ways to address the problem of unattended cooking (because it dominates the factors contributing to cooking fire ignitions) and scald safety (because it falls outside the scope of traditional fire safety), which quickly focused on ways to keep children away from danger zones where active cooking or hot food or drink might be located.

The EMAC messages were further processed by NFPA Public Education Division staff into a set of revised messages. In some cases, NFPA staff developed new messages independently to address gaps in the available messages. These messages are included in this report and displayed with the portion of the research that relates to them.

Chapter 1

. . . .

Cooking Fires and Injuries: The Size of the Problem

Cooking equipment has long been the leading cause of home fires and home fire injuries. When cooking equipment is described as a cause, it means that cooking equipment provided the heat that started the fire, not that the equipment malfunctioned. More cooking equipment fires are caused by human error than by malfunction. However, the equipment may have been less able to compensate for human error than other available equipment. For example, many coffee-makers and irons now shut off automatically after a period of time.

Cooking equipment was involved in 31 percent of the reported home structure fires in 2003.

NFPA estimates that cooking equipment was involved in 118,700, or 31 percent, of the home structure fires reported to U.S. fire departments in 2003.[2] (Homes include one- and two-family dwellings, apartments, and manufactured housing.) These fires caused an estimated 250 (8 percent) civilian deaths, 3,880 (29 percent) civilian injuries, and $512 million (9 percent) in direct property damage of the reported home fires and associated losses.*

For purposes of this analysis, cooking equipment is said to be involved if the incident type indicated a confined cooking fire or if the equipment involved was some type of heat-producing cooking equipment, a grease hood, or duct exhaust fan, or unclassified kitchen or cooking equipment.

NFIRS Version 5.0 introduced a "confined cooking fire" incident type code for fires involving contents of a cooking vessel without fire extension beyond the vessel.[3] The attraction of using the confined fire code option in NFIRS is that detailed information for this code is not required, although equipment involved was provided for about 10 percent of incidents reported as confined cooking fires. As a result, confined cooking fires accounted for 75,300, or 63 percent, of the 118,700 cooking fires reported in 2003.[2]

Confined cooking fires could have been coded in NFIRS Version 4.1 as fires with extent of flame damage coded as confined to object of origin. As more fires have been coded in NFIRS Version 5.0, the confined-fire percentage of estimated cooking fires has risen far past the percentage confined to object of origin in NFIRS Version 4.1. This and other patterns lead us to believe that many fires now coded as confined cooking fires would have

---

*Statistics extracted from Hall's 2006 report on cooking equipment fires exclude a share of the confined cooking fires based on the percentage of confined cooking fires with equipment information in which the equipment is not specifically intended for cooking, i.e., heating stoves. This analysis also excludes other types of kitchen equipment, such as refrigerators, dishwashers, blenders, and knives, which are not related to the process of heating food.

been considered smoke scares, and so not counted as fires, in NFIRS Version 4.1.

When including confined fires, cooking fires in 2003 were at the highest point since 1982.

Figure 3 shows that the 141,900 reported cooking fires in 1981 was the highest point since 1980, the first year of fire cause national estimates. Reported cooking fires hit their lowest point in 2000 with 93,700 such incidents. The increasing use of Version 5.0 of NFIRS has resulted in a growing number and share of confined cooking fires. Tracking trends has become more challenging with the introduction of confined fires in Version

5.0 of NFIRS in 1999. Because NFIRS requires only limited information on confined fires, they are easier to report. When the rules for data collection change, however, it is hard to discern whether increases are real or the result of changes in data collection practices. Figure 1 shows that the total of 118,700 cooking fires reported in 2003 is the highest since 1982. However, as noted above, it is possible that much of the increase since 1999 consists of confined cooking fires that would have been coded as smoke scares and not included in earlier estimates of cooking fires. If confined cooking fires are excluded for the years since NFIRS 5.0 was introduced, the number of cooking fires decreases dramatically.[2]

Figure 1. Reported Home Structure Cooking Equipment Fires in the U.S. by Year: 1980-2003

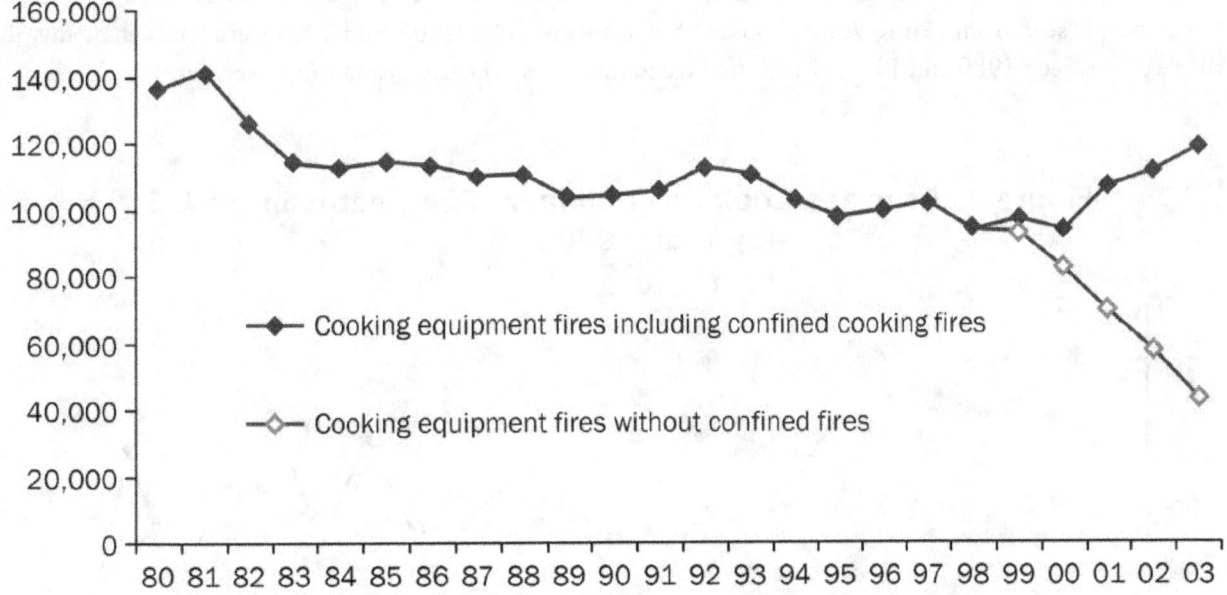

The share of reported home fires caused by cooking compared to other causes has increased over time.

Figure 2 shows that, in the 1980s, cooking equipment was involved in roughly one-fifth of the reported home structure fires. In the 1990s, this increased to one-quarter, and in 2003, it was close to one-third.[2]

Figure 2. U.S. Reported Home Structure Cooking Equipment Fires by Year and Percent of Total: 1980-2003

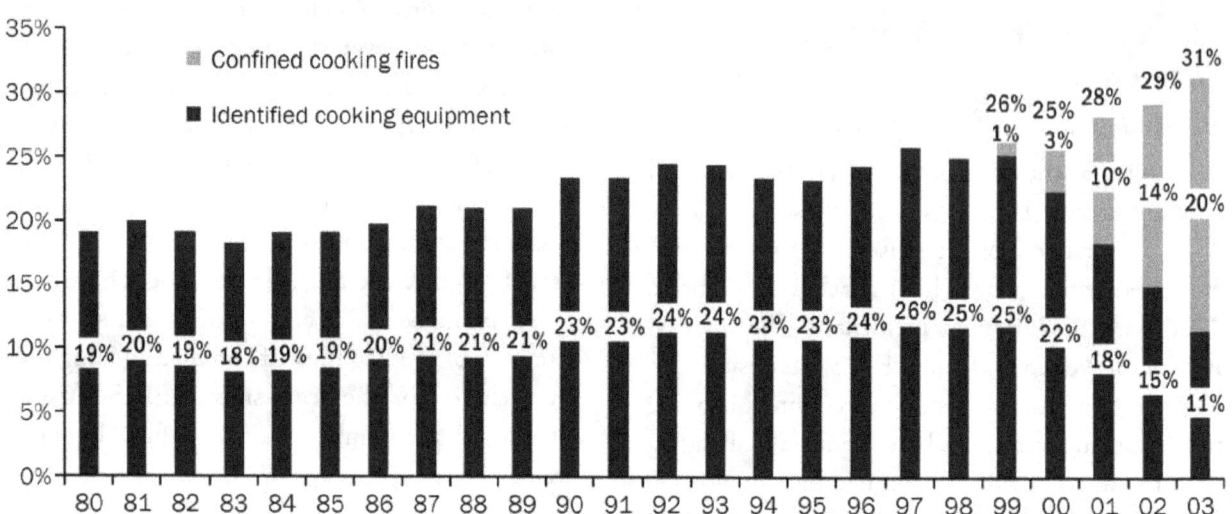

Cooking fire deaths have declined, but not consistently.

Although the trend in cooking fire deaths has generally been downward, Figure 3 shows that considerable fluctuation is seen from year to year.[2] Cooking fire deaths in 2000, 2002, and 2003 were lower than any of the years between 1980 and 1999. The dotted trend line shows the five-year annual averages.

Figure 3. Reported Cooking Equipment Fire Deaths in the U.S. by Year: 1980-2003

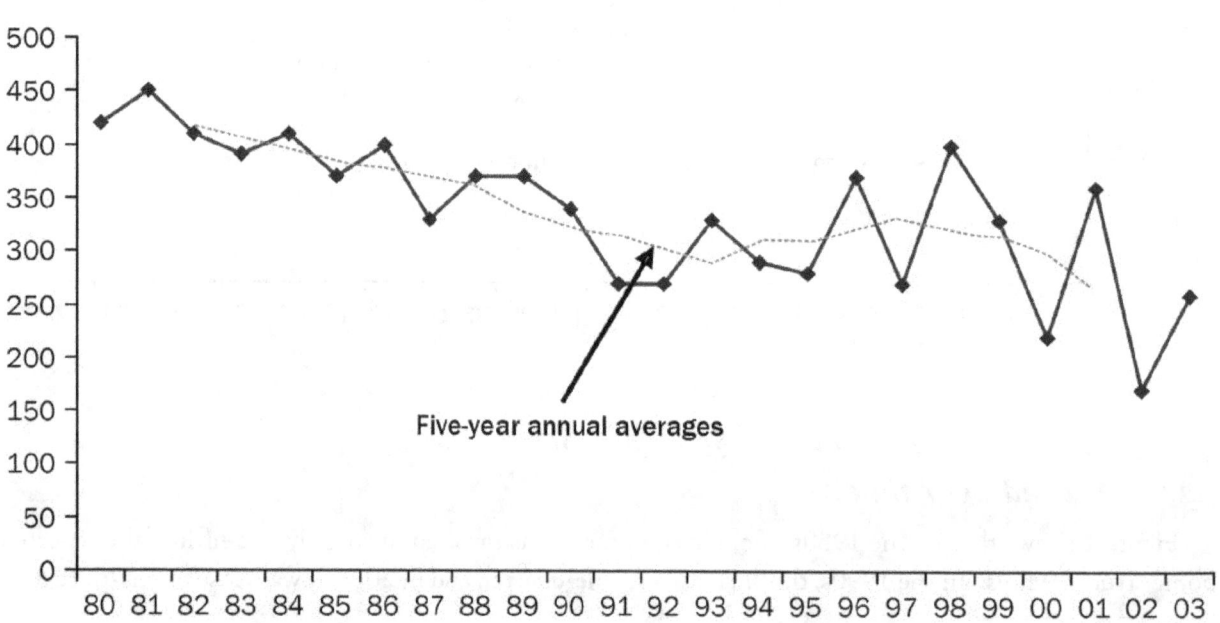

Reported cooking fire injuries hit their lowest point in 2002.

Figure 4 shows that reported cooking fire injuries hit their lowest points in 2001-2003, while these injuries peaked in 1993.[2] However, even with the record low numbers of injuries, the annual average of cooking fire injuries (including those from confined cooking fires) was only 12 percent lower than the annual average reported from 1980 to 1984. In addition, cooking is the leading cause of fire injuries.

Figure 4. Reported Cooking Equipment Fire Injuries in the U.S. by Year: 1980-2003

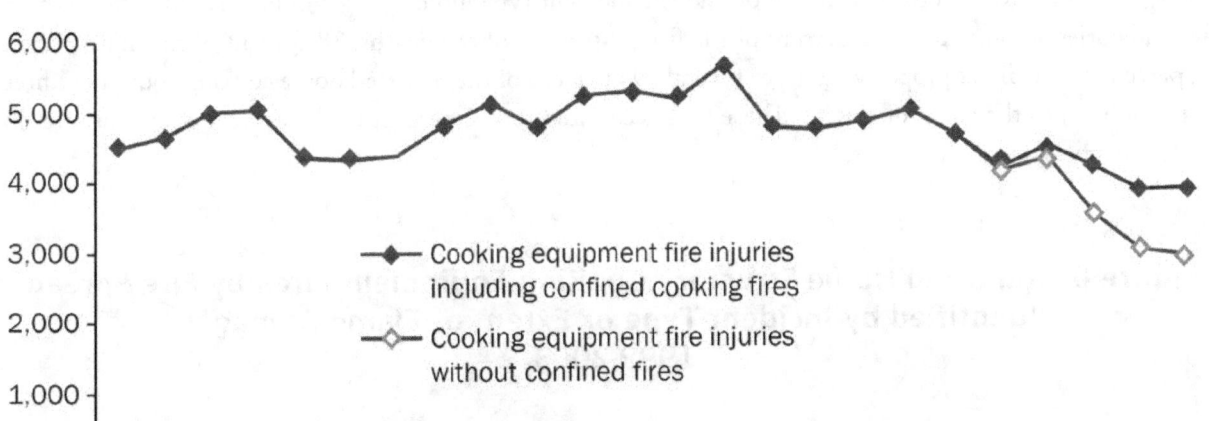

Figure 5. Reported Cooking Equipment Fires by Hour of Alarm: 1999-2003

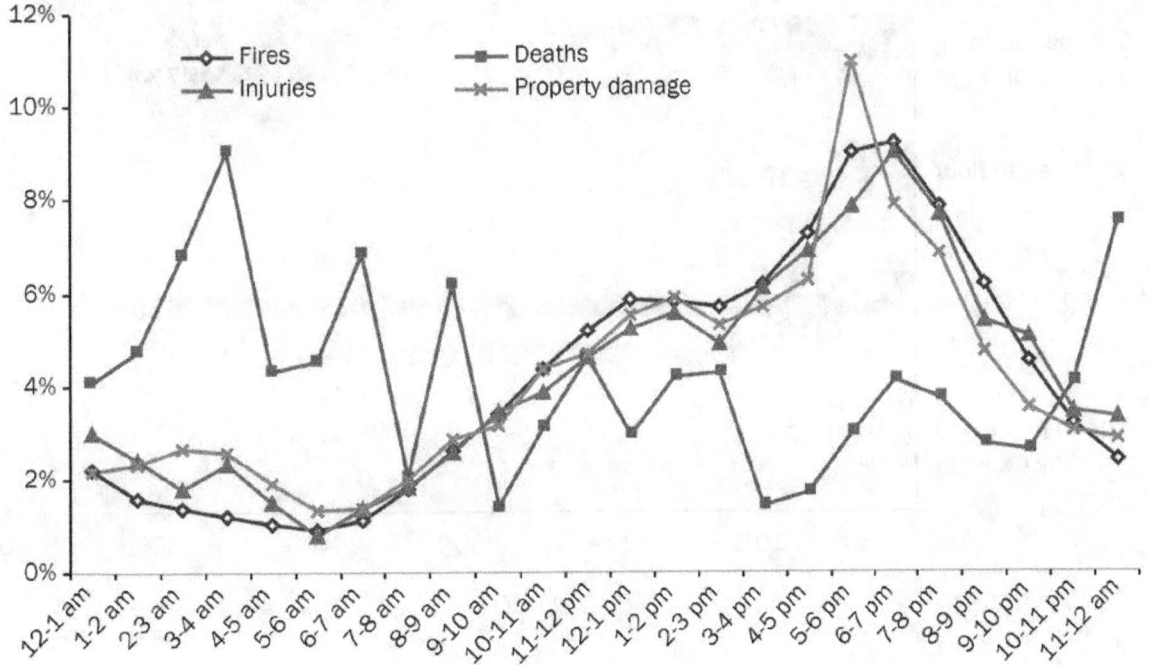

Cooking fires, injuries, and property damage peak around the dinner hour.

As Figure 5 shows, reported cooking fires, associated nonfatal injuries, and property damage follow very similar time patterns, climbing throughout the day and peaking between 5 p.m. and 7 p.m. The pattern for cooking fire deaths more closely resembles that seen for all fire deaths, with one-third of the deaths resulting from fires reported between 11 p.m. and 4 a.m.

Most reported cooking fires are small.

Figure 6 shows that more than two-thirds of the reported home structure cooking fires either had flame damage confined to the object of origin or had the incident type indicating a confined cooking fire. These two categories accounted for 71 percent of the fires, 8 percent of the deaths, 38 percent of the injuries, and 12 percent of the direct property damage. Overall, 95 percent of the reported home cooking equipment fires were confined to the room of origin. These fires accounted for 32 percent of the associated deaths and 85 percent of the associated injuries.

Figure 6. Reported Home Structure Cooking Equipment Fires by Fire Spread Identified by Incident Type or Extent of Flame Damage: 1999-2003

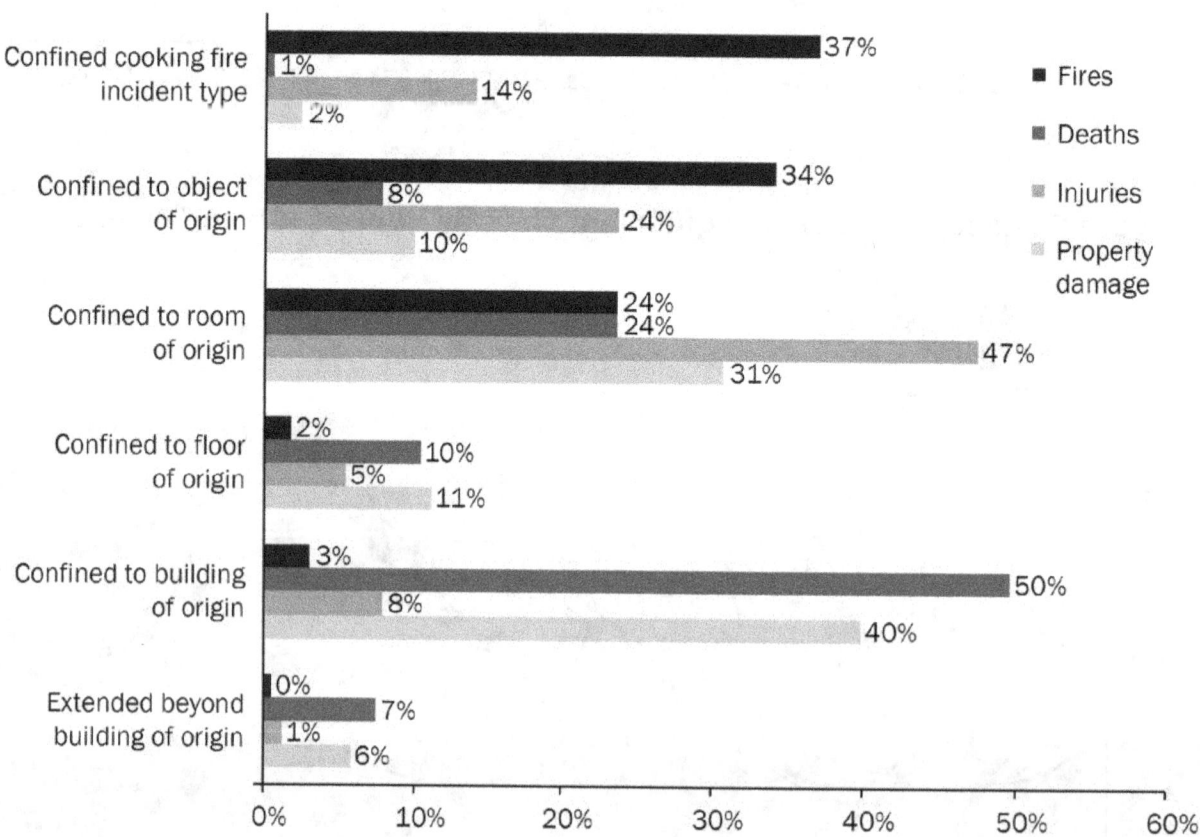

Most cooking fires are never reported to the fire department.

Based on a survey done for the CPSC from December 1983 to November 1984 using one- and three-month recall periods, it was estimated that kitchen or cooking equipment was involved in 12,244,000 unreported residential fires and 642,000 associated injuries or illnesses (headaches, dizziness, etc.).[4] This means that approximately 99 percent of all cooking fires are never reported to the fire department. Overall, 5 percent of unreported fires resulted in some type of injury or illness. Figure 7 shows that kitchen or cooking equipment was involved in 49 percent of the unreported fires in that study. An additional 19 percent were other kitchen fires. The same study estimated that only 4 percent of all types of residential fires are reported to fire departments.

Figure 7. CPSC's Unreported Residential Fires: December 1983-November 1984

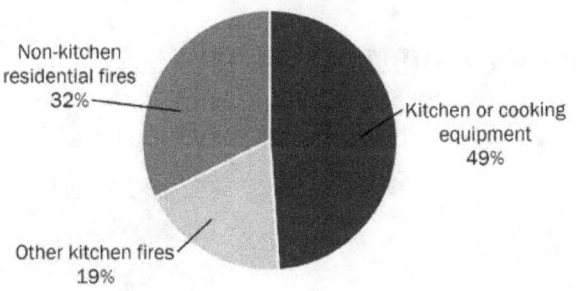

Many injuries seen at emergency rooms are not included in fire department reports.

CPSC used data from the NEISS to estimate the number of nonarson residential civilian fire injures treated in hospital emergency rooms from July 1, 2002, to June 30, 2003. Ovens or ranges were involved in 11,731, or 24 percent, of these injuries. Of the injuries involving ovens or ranges,

the fire service was in attendance at 6,560 (56 percent). The fire service also attended 370 (23 percent) of the 1,650 injuries involving grill fires.[5] These estimates of injuries resulting from reported cooking fires are higher than estimates developed by NFPA. In some cases, individuals may have been taken to the emergency rooms by private individuals or non-fire service agencies without fire department knowledge.

Summary Discussion

Due to the introduction of the confined cooking fire incident type in Version 5.0 of NFIRS, it is unclear whether cooking fires are actually increasing or decreasing. It is known, however, that cooking fire deaths and injuries have decreased since 1980. Regardless, cooking fires are still the leading cause of both reported and unreported home fires and home fire injuries. In addition, although the vast majority of cooking fires are minor and unreported, they still pose a significant risk of injury and death.

Implications for Behavioral Strategies

The cooking fire problem is sufficiently severe to warrant continued and, if possible, increased attention as a fire safety priority. The cooking fire problem's large share of total home fires and related civilian injuries suffice to make that case. As a result, it is imperative that the fire service community continue to educate people about and urge them to practice safe cooking behaviors.

Chapter 2

• •

Characteristics of Cooks and People Injured in Cooking Fires

To prevent cooking fires, it is necessary to know who is cooking and who is at risk from cooking fires. Social, environmental, and personal factors such as presence of distractions when cooking, age, time pressure, clutter, use of alcohol or medication, and mobility or agility can increase or decrease the risk of a cooking fire or injury.

While women spend more time on cooking-related activities, more males died from home cooking fires from 1999 to 2003.

U.S. women at least 15 years of age spend an average of 47.4 minutes a day on food preparation and cleanup in a typical day. Men, on the other hand, spend an average of 15 minutes a day on these same tasks.[6] However, Figure 8 shows that, from 1999 to 2003, males accounted for 56 percent of the home cooking fire deaths and 47 percent of cooking fire injuries. Considering that men spend one-third of the time that women spend on food preparation and cleanup, the male risk from these fires is substantially higher.

Figure 8. Cooking Equipment Fire Victims by Gender

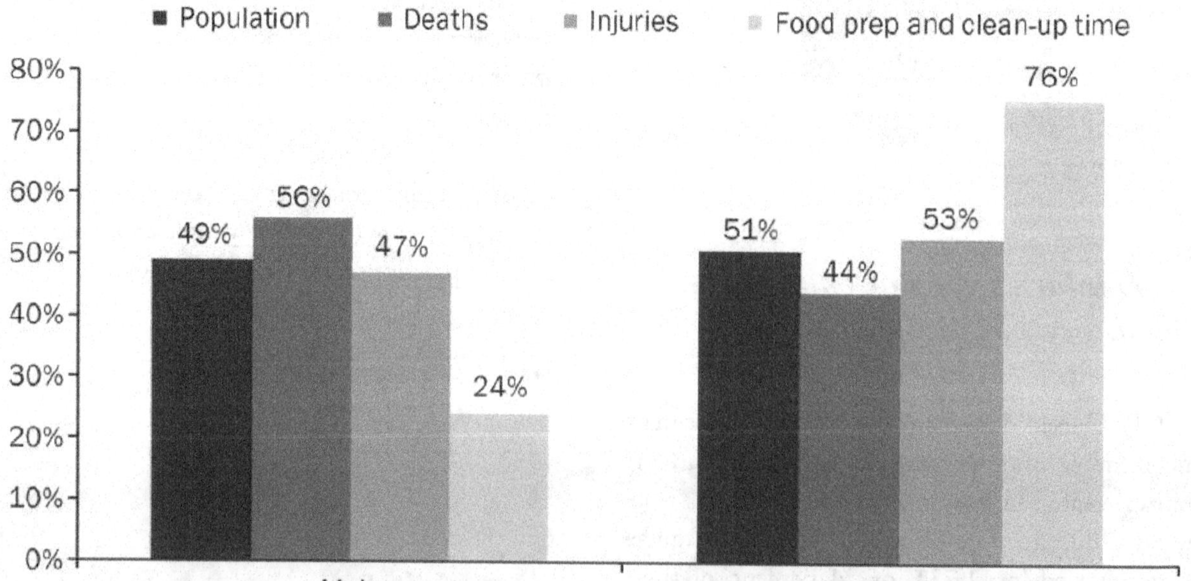

The cook in most cooking fires was an older teen or an adult under 70 years of age.

A 1995-1996 study of reported cooking fires in 10 communities done by the National Association of State Fire Marshals (NASFM) Cooking Fires Task Force and Association of Home Appliance Manufacturers (AHAM) Safe Cooking Campaign asked about the age of the cook involved in the cooking fire. Individuals between ages 19 and 69 faced a disproportionate risk of cooking fires compared to their share in the general population. The risk was highest for those between 30 and 49.[7] As part of

a different special study of range fires, CPSC analyzed the results of 289 field investigations of fire service-attended range fires that occurred between October 1994 and July 1995. Figure 9 shows that 84 percent of the cooks in these fires were between ages 15 and 64, with those between 15 and 44 years of age having a range fire risk of roughly 1.5 times that of the general population. This is based on the number of fires, the age of the cook, and the percentage of the population in the different age groups.[8] Unfortunately, no data were found on age differences in time spent cooking or in the number of meals prepared.

Figure 9. Age of Cook in Food Ignitions: CPSC Range Fire Study

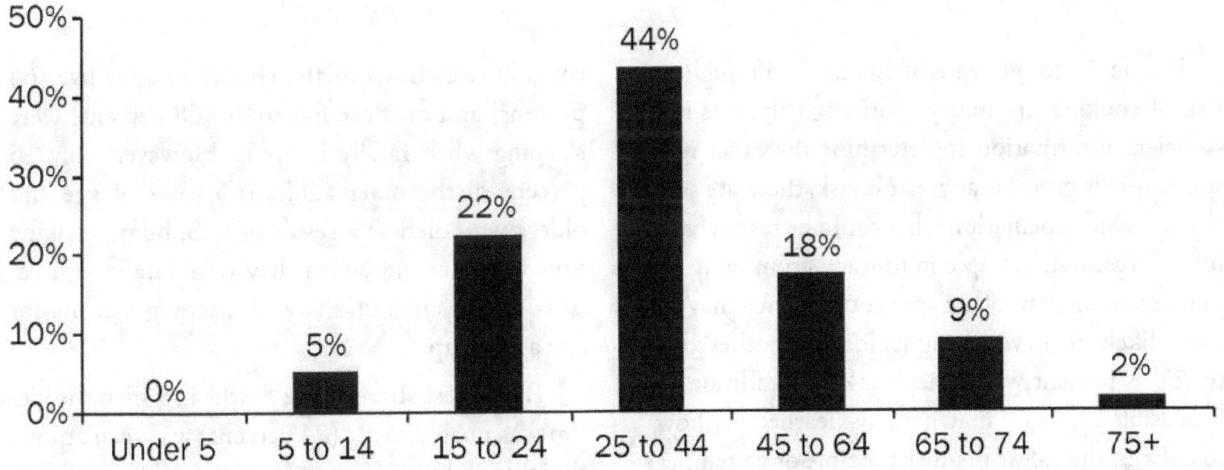

Older adults and very young children account for a disproportionate share of cooking fire deaths.

Two-thirds (67 percent) of the population is between 15 and 64 years of age. This age group accounts for half (52 percent) of the cooking fire deaths and three-quarter (76 percent) of the cooking fire injuries. Figure 10 shows, however, that while only 12 percent of the U.S. population is 65 years of age or older, these individuals accounted for 30 percent of the cooking fire deaths. In addition,

while only seven percent of the population is under five years of age, this group accounted for nine percent of the cooking fire deaths.[2] These statistics are based on all victims, not just the cooks.

Young children and older adults are at higher risk for death in home cooking fires, but to roughly the same extent that they are at higher risk for death in most types of home fires. This may mean that the higher risk is less a matter of the special difficulties they have in cooking and more a matter of the special difficulties they have in responding to fires.

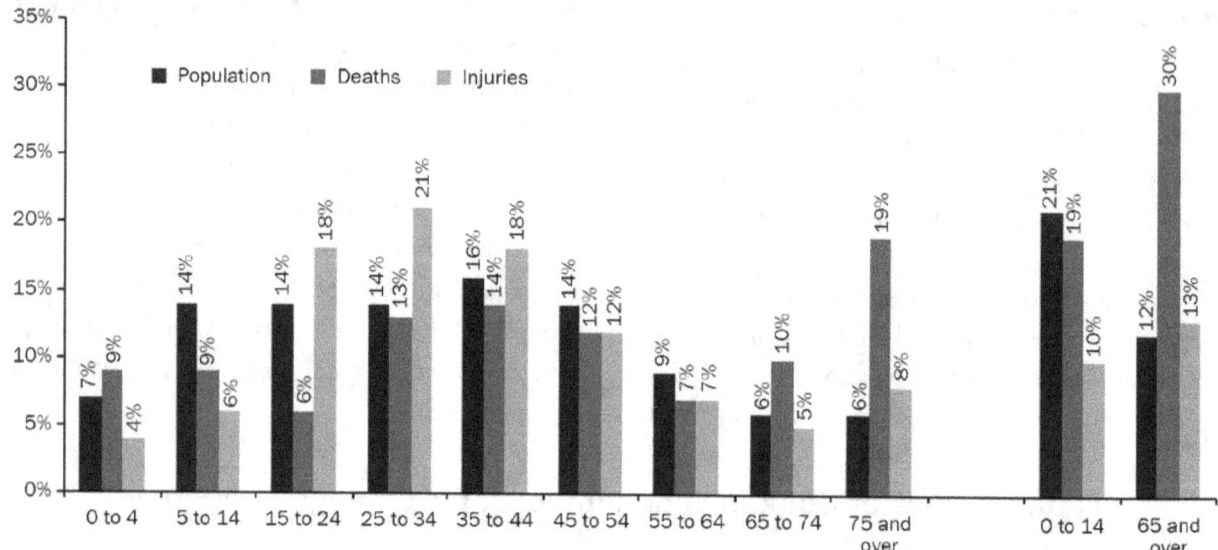

Figure 10. Percent of Home Cooking Equipment Fire Deaths and Injuries Compared to Population, by Age Group: 1999-2003

People 25 to 34 years of age faced the highest risk of cooking fire injury. Although there is not sufficient information to determine the exact reason why this group is at greatest risk, there are several possible explanations that could be tested with further research. People in this age group may do more cooking than other age groups. They may be more likely to have young children or other distractions present when they cook. In addition, it is possible that they may not have learned yet how to cook in the safest manner possible or to temper their boldness in all things with a caution born of an awareness of their mortality. Youths and young adults 15 to 24 years of age, adults aged 35 to 44, and people 75 years of age or older also faced an elevated risk of cooking fire injuries.

Sleeping was the most common activity among civilians who were fatally injured in cooking fires.

The leading activity at time of injury varies between fatal and nonfatal cooking fire injury and between age groups. Figure 11 shows almost half (46 percent) of the adults ages 25 to 64, and

roughly two-thirds of the children under five (64 percent) and of those five to 24 (68 percent) were sleeping when fatally injured. However, only 15 percent of the older adults (65 years of age and older) who died as a result of U.S. home cooking fires were sleeping when they were fatally injured. This is the smallest share of sleeping victims for any age group.

The largest share of fire deaths in which the victim was unable to act (24 percent) was seen among the older adults. Twenty-two percent of children under five who died from cooking fires also were described as unable to act. This description may be a reflection of physical disabilities that sometimes accompany an advanced or very young age.

Firefighting was the most common activity among civilians who were nonfatally injured in cooking fires and were over 5-years old.

Figure 12 shows that 44 percent of the injuries incurred by those 65 years of age and older, 55 percent for those 5 to 24 years of age, and 60 percent for those 25 to 64 years of age were incurred

Figure 11. Home Cooking Equipment Fire Deaths by Activity at Time of Injury and Age Group: 1999-2003

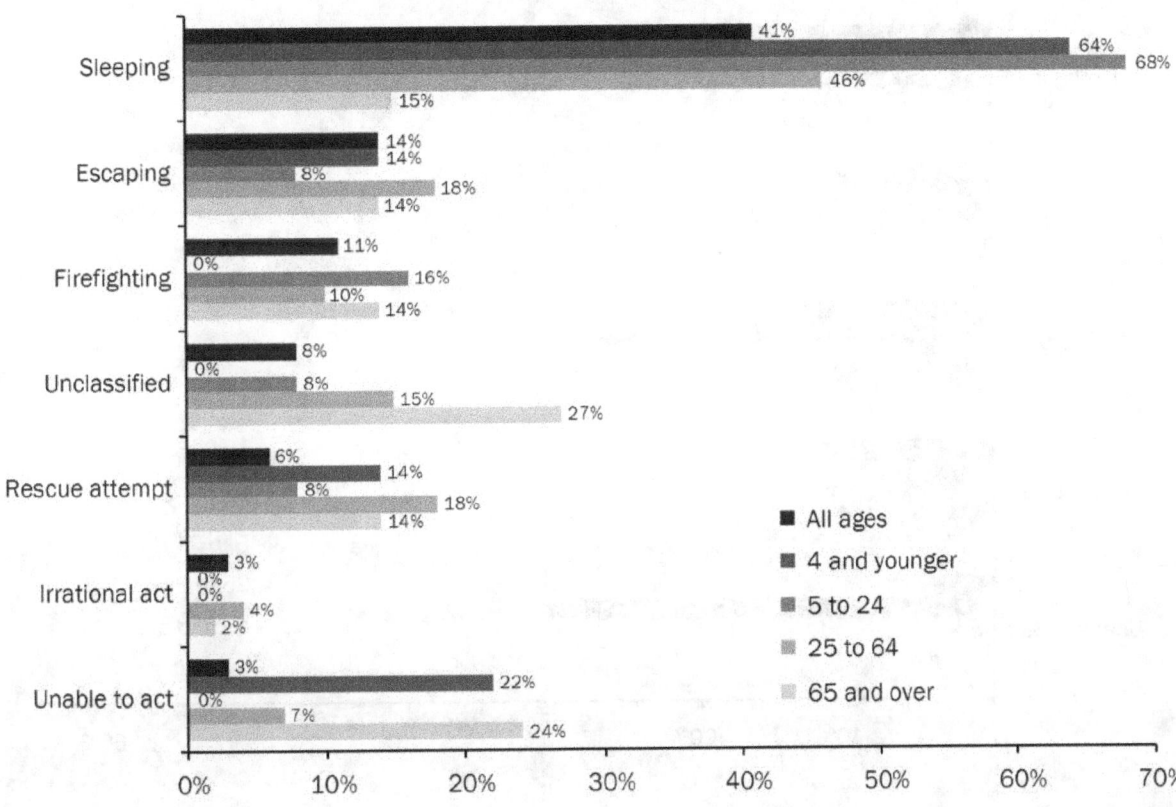

while fighting the fire. Chapter 5 provides a more detailed examination of civilian firefighting with regard to home cooking fires.

Little gender difference is seen in cooking fire activities when injured.

Figure 13 shows that little difference is seen in the gender patterns in activity when non-fatal cooking injuries were incurred in home cooking fires. Fifty-six percent of the males and 54 percent of the females were attempting to fight the fire when injured. Fourteen percent of the females and 10 percent of the males were injured while escaping.

People have different levels of interest in cooking.

Different types of stove users were identified in a course project at George Mason University.[9]

For example, someone like a conventional home-maker does the majority of cooking for family meals and bakes often. Equipment that is easy to use in terms of pre-heating, baking, broiling, boiling, and simmering is important to this individual who carefully follows recipes received from friends and magazines. Such an individual may be more likely to read women's magazine than magazines specifically about cooking. Other individuals who are more interested in innovative cooking often try new techniques and tools to prepare gourmet meals. These individuals tend to improvise on recipes and are more likely to watch cooking shows on television and buy gourmet publications. A third group wants very basic cooking equipment as they use the stove and microwave primarily to heat food, rather than to prepare it. These individuals may be less likely to be interested in reading or watching anything specifically about cooking.

Figure 12. Nonfatal Home Cooking Equipment Fire Injuries by Leading Activities at Time of Injury and Age Group: 1999-2003

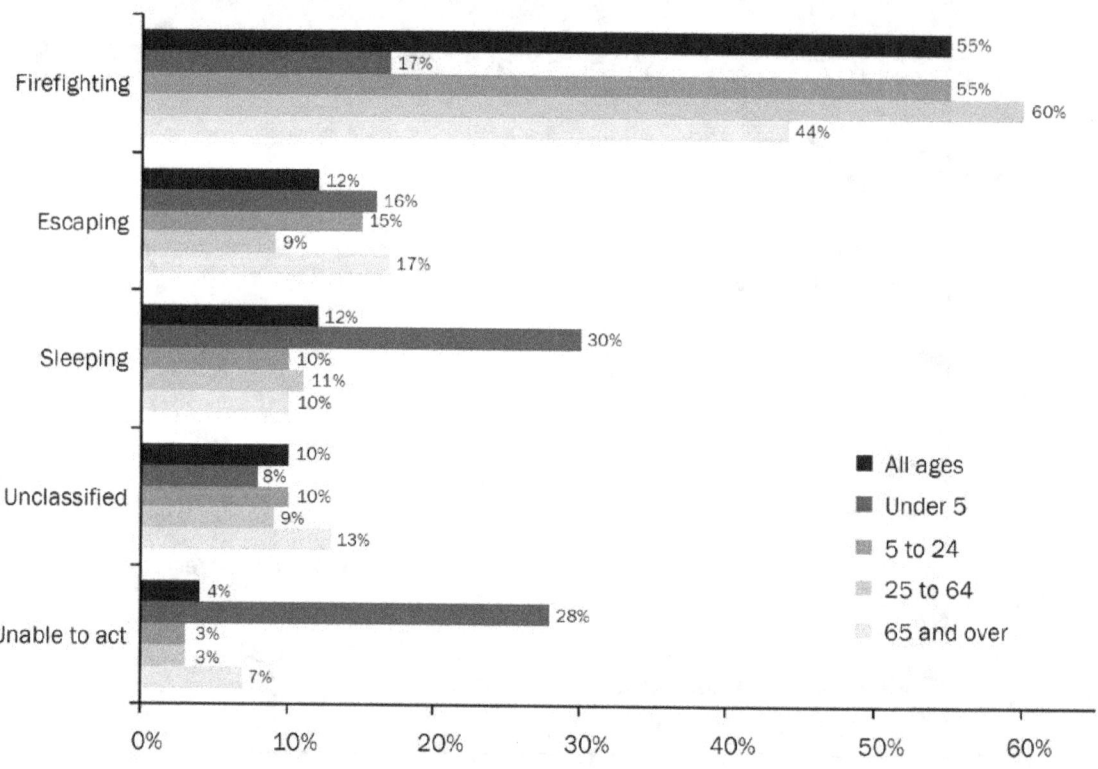

Figure 13. Home Cooking Equipment Fire Injuries by Activity at Time of Injury and Gender: 1999-2003

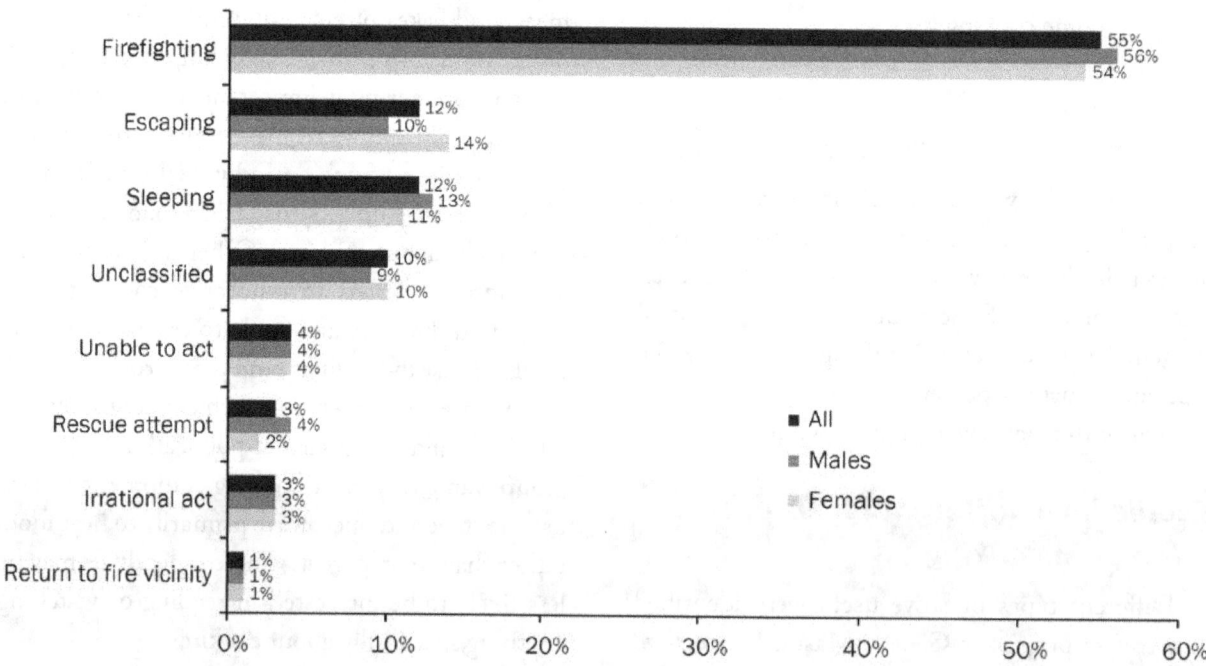

The interests and preferences of the cook influence the type of cooking, some of the risks that might be taken, and perhaps the best venues for communicating safety information.

Summary Discussion

Although women do most of the cooking, males face a disproportionate risk of cooking fire injury and death relative to the time spent cooking. The majority of cooks in cooking fire studies were either older teens or adults under 70 years of age. However, in terms of the total population, adults 65 years of age and over and children under five accounted for a disproportionate share of cooking fire deaths. Individuals between ages 25 and 34 faced the highest risk of cooking fire injuries. Youths and young adults 15 to 24 years of age, adults aged 35 to 44, and those 75 years of age or older also faced an elevated risk of cooking fire injuries. The leading activity at time of fatal injury was sleeping for all age groups except for older adults. The leading activity at time of nonfatal injury was firefighting for all age groups studied except for children under five.

Implications for Behavioral Strategies

These demographics become important when developing cooking safety messages and determining the most appropriate venues for these messages. The challenge is to develop and publicize materials that will be relevant to the different at-risk groups identified through research (women versus men, older adults, etc.), recognizing that different groups may respond better to different emphases. For example, women's magazines may reach many of the cooks, but are unlikely to be read by men. Also, given the higher injury rate among teens and young adults, additional efforts should be made to reach that population. Finally, while many people enjoy cooking, there are others who consider it a chore and would have little interest in any material on the topic. As a result, careful consideration must be given to how the fire service community can spread safety messages effectively to different groups of people.

Chapter 3

. . . .

Patterns by Type of Cooking Equipment

The frequency of reported cooking fires varies by type of cooking equipment. In addition, the quality of equipment (how well it is maintained and initially made or installed) factors into the likelihood of fire.

Ranges* are the leading type of cooking equipment involved in fires.

From 1999 to 2003, ranges were involved in two-thirds of the reported home cooking fires (67 percent) and four-fifths of the associated civilian deaths (82 percent) and injuries (80 percent). Range fires also caused roughly two-thirds (67 percent) of the cooking fire direct property damage. Both confined and nonconfined fires are included. During this time period, when incidents coded as confined cooking fires had equipment involved, ranges were involved in 53 percent of the fires and ovens in 23 percent of the fires.[2]

In addition, ranges or stoves accounted for 49 percent of the kitchen or cooking equipment fires in CPSC's study of unreported residential fires.[4] However, although ranges and stoves are still the leading equipment type, the ratio of range and stove fires to oven fires is substantially lower for unreported cooking fires than for total fires reported to the fire departments and is closer to the ratio for confined cooking equipment fires from 1999 to 2003 with identified equipment.[2] This means that the smaller the cooking fire, the more likely it is to be an oven fire as opposed to a range or stove fire.

Only 12 percent of reported U.S. home cooking fires were attributed to equipment failures.

Overall, equipment failures caused only 12 percent of the reported home cooking equipment structure fires from 1999 to 2003, 8 percent of the associated civilian deaths, 7 percent of the associated injuries, and 11 percent of the associated direct property damage. Figure 14 shows that the percentage of fires resulting from equipment failure varies considerably by device. Microwave ovens, grease hoods or ducts, and gas grills make up the largest share of such fires. Grease hoods and ducts function with less human interaction in comparison with the other devices. As a result, the high share for equipment-related factors is not surprising for them.[2]

Electrical problems are more common with electric ranges and ovens than with gas-fueled equipment.

Fifty-nine percent of U.S. households cooked with electricity in 2003.[10] Including adjustments

---

* While a separate NFIRS code exists for ovens and rotisseries, the range category includes ranges with and without ovens as well as cooktops only. As a result, range fires are likely to include some incidents that began in the oven portion of the range.

Figure 14. Percent of Home Cooking Equipment Structure Fires Caused by Equipment Failure: 1999-2003

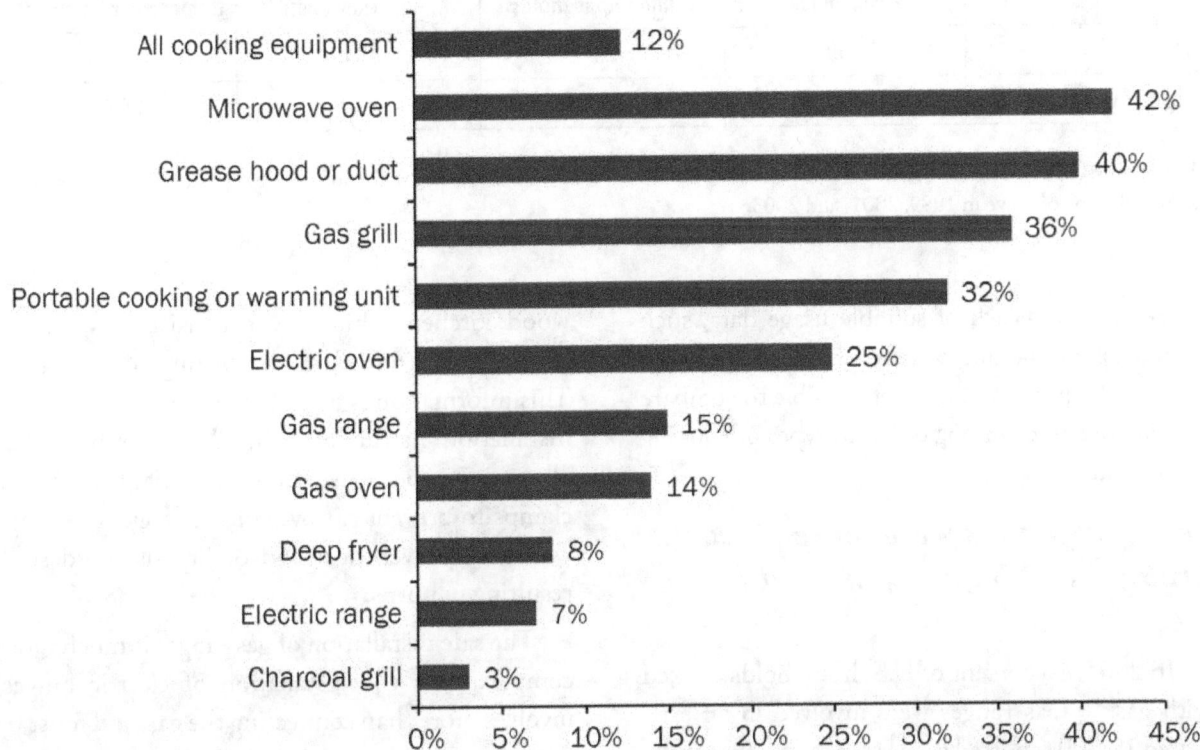

for confined fires,* electric ranges were involved in an estimated 58,200 reported home structure fires. These fires caused 100 civilian deaths, 2,490 civilian injuries, and $266 million in direct property damage. Electric ovens were involved in an estimated 15,900 reported home structure fires, resulting in 11 civilian deaths, 290 civilian injuries, and $37 million in direct property damage. These figures also include adjustments for confined fires.

From 1999 to 2003, short circuit arcs or other electrical failures or malfunctions were factors in 29 percent of the microwave oven fires and 17 percent of the portable cooking equipment fires. Although

not a significant factor in fires involving gas ranges and ovens, short circuit arcs or other electrical failures or malfunctions were factors in 5 percent of the electric range fires and 15 percent of the electric oven fires.[2]

Table 1 shows that, based on the number of households cooking with electric or gas stoves, the risk of fire per million households was 47 percent higher for electric stoves in terms of reported fires. In addition, the risk of reported civilian injury and property damage was more than twice as high for electric stoves. On the other hand, the risk of fire death was 15 percent higher for gas stoves than electric.[2]

---

* Because causal information is not required for confined fires, the number of specific types of equipment reported as involved in fires declined sharply. Specific equipment information was provided in roughly 10 percent of the confined cooking fires, making it possible to use this information to calculate the percentage of confined fires in which specific types of equipment were involved or in which specific causal factors occurred or were present. These percentages then are applied to the total confined fires, and the resulting statistics are added to the nonconfined fire totals.

Table 1. Fire Risk for Electric and Gas Stoves Based on Households Using Each*

	Fires per Million Households	Civilian Deaths per Million Households	Civilian Injuries per Million Households	Direct Property Damage per Household
Gas	359	2.2	15.6	$1.73
Electricity	528	1.9	34.1	$4.03

*Based on national estimates of home structure fires involving each from 1999 to 2003 and the average number of households using each type of stove in 1999, 2001, and 2003.

Because of a lack of suitable usage data, such as time spent cooking or percent of meals cooked by type of equipment, it is not possible to compare cooking fire risks among different types of cooking equipment.

Leaks or breaks were more frequent problems with gas equipment than electric.

In 2003, 40 percent of U.S. households cooked with gas.[10] Gas ranges were involved in an estimated 19,500 reported home structure fires, resulting in 100 civilian deaths, 530 civilian injuries, and $66 million in direct property damage. Gas ovens were involved in an estimated 7,600 reported home structure fires, resulting in 90 civilian injuries, and $15 million in direct property damage. No deaths from gas oven fires were reported in 2003. These statistics also include adjustments for confined fires. While not a significant factor in electric range and oven fires, leaks or breaks were factors in 13 percent of the gas range fires and 6 percent of the gas oven fires from 1999 to 2003.[2]

Choose approved equipment and follow instructions for installation and use to prevent fires.

First, it is important always to use cooking equipment tested and approved by a recognized testing facility. Second, installers of tested and approved gas and electric stoves must verify that the clearances to combustible materials, such as

wood kitchen cabinets, are consistent with the distance determined in the testing of the range. This information is included in the manufacturer's installation instructions. In addition, ranges usually require the installation of special brackets or clamps to prevent tip-over of the range. If a range were to tip over, hot food or liquids could spill, resulting in burns.

The safe installation of gas ranges is much more complex than the installation of electric ranges, involves more than connecting the gas, and must be done in accordance with installation requirements and codes. Signs of problems with gas ranges are flames that are yellow, uneven, or that "float" above the burner. In addition, leaks can develop at anytime in the valves incorporated in ranges. An important signal of leaks is the distinctive gas odor that is especially detectable upon entering the building from the outside.

Finally, the air flow in a gas oven should never be blocked as this may cause inadequate operation and even carbon monoxide poisoning. Any slots, holes or passages in the bottom of a gas oven should always be kept clear and the entire rack should never be covered with materials, such as aluminum foil, that could trap heat causing a fire hazard.[11]

Cooking fire safety rules need to be tailored to equipment differences.

Traditional messages caution against leaving the room when cooking with any type of equipment. However, cooking safety messages should be relevant to the types of equipment used. For example, slow cookers are designed to operate safely without constant attention.

Serious home cooks often seek equipment not traditionally associated with the home that may require special consideration.

Consumers sometimes purchase equipment originally designed for restaurants or caterers, such as ranges, butane-fueled tabletop burners, turkey fryers, and crème brulée torches.

+ Home ranges are tested to different standards than restaurant ranges.

Home ranges usually are tested to verify that sides and backs will not get hot enough to ignite wood kitchen cabinets and other combustible materials. Restaurant ranges, however, are not required to meet the same criteria. Consequently, a few inches of clearance (open space) may be needed from combustible materials. Home ranges also are tested to ensure surfaces and handles will not get hot enough to cause burns. Homeowners wishing to install a restaurant or commercial-type range should purchase a commercial-type range designed and tested for household use.[11]

+ Many are concerned by the increasing popularity of turkey fryers in recent years.

NFPA strongly discourages the use of turkey fryers except by properly trained professionals using professional-quality equipment.[12] Turkey fryers use a substantial quantity of cooking oil at high temperatures. Units currently available for home use pose a significant danger that hot oil will be released at some point during the cooking process. The risks of tip over, splashing, spilling, fire, or rain or moisture coming into contact with the 5 gallons of hot oil are seen as too high by NFPA and Underwriters Laboratories, Inc (UL). As a result, UL has decided not to certify

these fryers.[13] Some fire departments, however, believe that these fryers will be used regardless and issue guidelines for safe use.[14,15]

+ Importing portable butane stoves has increased significantly.

While existing standards address commercial butane-fueled tabletop cooking appliances and portable gas camp stoves, these appliances also have been marked for home use by consumers.

The CPSC conducted indepth investigations into 14 incidents involving such products that occurred between January 1, 1995, and August 21, 2001. The design in question included a disposal 8-ounce butane canister that fits alongside the burner. Twenty-four injuries resulted from these incidents. Failures in the fuel compartments were noted in all 14 investigations. Fire was reported in 12 of the 14 incidents and 21 of the 24 injuries. Three injuries resulted from hot food and broken dishes associated with sudden pressure release in two incidents. The incidents occurred in both commercial and noncommercial occupancies, indoors and outside.

Two overheating scenarios were identified. In some cases, large pans extended over the fuel canister and restricted the air flow. In models of older design, the drip pan and grate had been inverted for shipping to save space. In four incidents, the drip pan was still inverted. One user assumed the equipment was shipped the way it should be used. In another four incidents, two of these devices were used right next to each other. Some manufacturers caution against this because of the increased heat exposure to the butane canisters. In the typical injury scenario reported, the appliance had been operating for at least 5 minutes when the user saw an explosion and flames, sometimes shooting as high as 6 feet.

The CPSC has identified the following three main issues with butane-fueled tabletop cooking appliances: (1) overpressure protection is not required by U.S. voluntary standards; (2) consumers tend to use the device configured as they were originally packaged, even if the grate is incorrectly packaged upside down; and (3) the scope of the standards is limited to outdoor and commercial use. CPSC recommends that: (1) voluntary standards incorporate overpressure protection performance criteria similar to those found in the Japanese and Korean standards; (2) either the stoves be required to be usable safely as packaged, or interlock be required ensuring the grate and drip pan are in proper position before the fuel flows; and (3) the scope of the standards be expanded to include household use.[16]

Aluminum pans contribute to the fire and burn problem.

The CPSC warns that empty or almost empty aluminum cookware (or steel cookware with an aluminum core) on high heat can "boil dry." If such a pan is picked up, molten aluminum can drip and cause burns. Overheated aluminum cookware also can cause fires. Such cookware should not be preheated on high heat. Should such a pan boil dry and start to melt, consumers are advised to shut the heat off and leave the pan in place until it cools.[17]

Grilling and Outdoor Cooking Fires and Fire Safety

Although most family cooking is done in the kitchen, a considerable portion is done outside on barbecue grills. While many of the same kitchen cooking precautions apply to grilling, some aspects of outdoor cooking require special care and should be carried out in designated areas.

Exterior balconies or unenclosed porches were the leading area of origin for home gas and charcoal grill structure fires.

From 1999 to 2003, an exterior balcony or unenclosed porch was the area of origin in 32 percent of the gas grill home structure fires and 45 percent of the home structure fires started by charcoal grills. This area also may include decks.[2]

Leaks or breaks and combustibles too close to the heat source were leading factors in grill fires.

In 2003, with adjustments for confined fires, gas grills were involved in an estimated 900 home structure fires and 2,500 outside or unclassified fires on home properties. Leaks or breaks contributed to 30 percent of these structure fires and 46 percent of these outdoor fires. With similar adjustments for confined fires, charcoal grills were involved in 600 home structure fires and 300 outside or unclassified fires in the same year. Combustible too close to the heat source was the leading factor in charcoal grill fires. This factor was also the second leading cause of home structure fires started by gas grills.[2]

Summary Discussion

Ranges dominate the cooking fire problem, and both gas-fueled and electric-powered ranges contribute to or are involved in a significant numbers of fires. The risk of reported fire, injury, and property damage was higher from electrical stoves than from gas, while the risk of fire death was higher from gas stoves. Leaks or breaks are more common factors in gas-fueled equipment than in electrical, while short circuits and electrical failures are more common in electrical cooking equipment than in gas. As neither power type poses a consistently higher risk of cooking fires on all measures of loss, behavioral strategies need to address both types of appliances. In addition, the fire service community and cooking equipment manufacturers must ensure that people know that they have a responsibility to install all cooking equipment in accordance with installation requirements, be alert and mindful of leaks or mechanical problems that could happen at anytime, and operate the equipment as safely as possible at all times.

As the levels of interest in cooking and types of practices constantly change, the use of different types of specialized cooking equipment increases. As each additional piece of specialized cooking equipment poses its own unique risks to the practice of cooking, it is important for the fire service community to promote behavioral mitigation messages specific to these specialized types of equipment and associated behaviors.

Outdoor grilling involves a number of distinct safety issues. Because serious fire loss is extremely rare in the absence of structural involvement, the safety issue that must be given the highest priority is positioning the grill away from all structures. Gas and charcoal grills have different safety requirements based on the fuel used.

Behavioral Strategies

The following specific messages arising from this chapter address choosing the right equipment and using it properly:

- Always use cooking equipment tested and approved by a recognized testing facility.

- Follow manufacturers' instructions and code requirements when installing and operating cooking equipment.

- Plug microwave ovens and other cooking appliances directly into an outlet. Never use an extension cord for a cooking appliance, as it can overload the circuit and cause a fire.

The following general cooking messages have been adapted to apply to outdoor grilling, with separate messages for all types of outdoor grills and messages for charcoal and gas grills respectively:[18]

1. Using barbecue grills safely

   - Position the grill well away from siding, deck railings, and out from under eaves and overhanging branches.

   - Place the grill a safe distance from lawn games, play areas, and foot traffic.

   - Keep children and pets away from the grill area by declaring a 3-foot "kid-free zone" around the grill.

   - Put out several long-handled grilling tools to give the chef plenty of clearance from heat and flames when cooking food.

   - Periodically remove grease or fat buildup in trays below grill so it cannot be ignited by a hot grill.

   - Use only outdoors! If used indoors, or in any enclosed spaces, such as tents, barbecue grills pose both a fire hazard and the risk of exposing occupants to carbon monoxide.

2. Charcoal grills

   - Purchase the proper starter fluid and store out of reach of children and away from heat sources.

   - Never add charcoal starter fluid when coals or kindling have already been ignited, and never use any flammable or combustible liquid other than charcoal starter fluid to get the fire going.

3. Propane grills

   - Check the propane cylinder hose for leaks before using it for the first time each year. A light soap and water solution applied to the hose will reveal escaping propane quickly by releasing bubbles.

   - If you determined your grill has a gas leak by smell or the soapy bubble test and there is no flame:

     - Turn off the propane tank and grill.

     - If the leak stops, get the grill serviced by a professional before using it again.

     - If the leak does not stop, call the fire department.

+ If you smell gas while cooking, immediately get away from the grill and call the fire department. Do not attempt to move the grill.

+ All propane cylinders manufactured after April 2002 must have overfill protection devices (OPD). OPDs shut off the flow of propane before capacity is reached, limiting the potential for release of propane gas if the cylinder heats up. OPDs are easily identified by their triangular-shaped hand wheel.

+ Use only equipment bearing the mark of an independent testing laboratory. Follow the manufacturers' instructions on how to set up the grill and maintain it.

+ Never store propane cylinders in buildings or garages. If you store a gas grill inside during the winter, disconnect the cylinder and leave it outside.

The American Burn Association included camping burn prevention in its 2002 Burn Awareness Week.

In 2002, Summer Recreational and Camping Burn Prevention was the theme of the American Burn Association's (ABA's) Burn Awareness Week.[19] Their materials included extensive safety tips on outdoor cooking and grilling. While many of these messages are quite similar, some are more comprehensive than those listed here. These messages are found in Appendix C.

Chapter 4

• •

Behaviors and Types of Cooking Associated with Cooking Fires

Behavioral factors, such as the amount of attention paid to the cooking and separating combustibles from the heat source, and types of cooking, such as frying and boiling, also affect the likelihood of having a cooking fire.

Unattended equipment was the leading factor contributing to home cooking fires.

From 1999 to 2003, cooking equipment had been left unattended in 37 percent of the home

Figure 15. Leading Factors Contributing to Ignition in U.S. Home Cooking Fires: 1999-2003

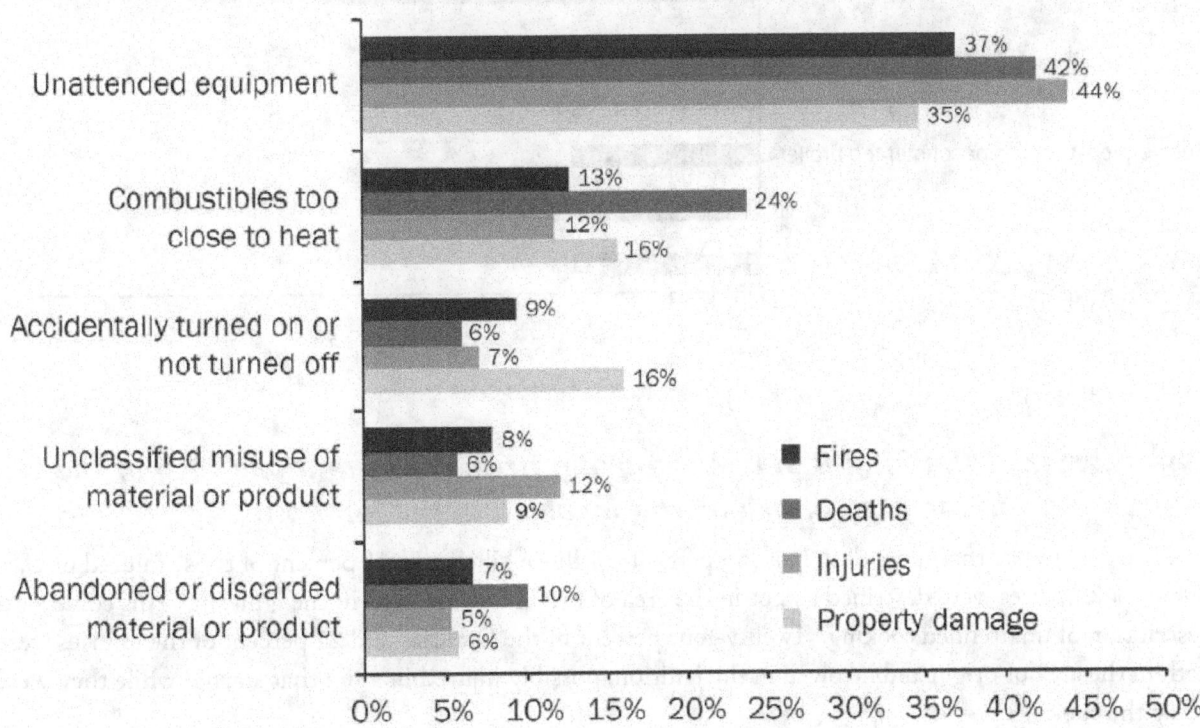

cooking equipment fires reported in Version 5.0 of NFIRS.* Figure 15 shows that unattended equipment was also a factor in 42 percent of the cooking fire deaths and 44 percent of the injuries. These statistics may be even be higher, as it is possible that the 7 percent coded as abandoned or discarded material or product may also represent unattended cooking. Some type of combustible material too close to the cooking equipment was a factor in 13 percent of home cooking fires, 24 percent of the associated deaths, and 12 percent of the associated injuries. Unintentionally turning the equipment on or failing to turn it off was a factor in 9 percent of the fires.

The share of fires resulting from unattended equipment varied by the type of cooking equipment involved. While unattended equipment was a contributing factor in 37 percent of the reported cooking fires overall, Figure 16 shows that it was a factor in 45 percent of the deep fryer fires and 43 percent of the range fires. It was cited as a factor in only 21 percent of the conventional oven or rotisserie fires and 17 percent of the microwave oven fires.

Figure 16. Unattended as Percent of Home Cooking Structure Fires by Leading Equipment Types: 1999-2003

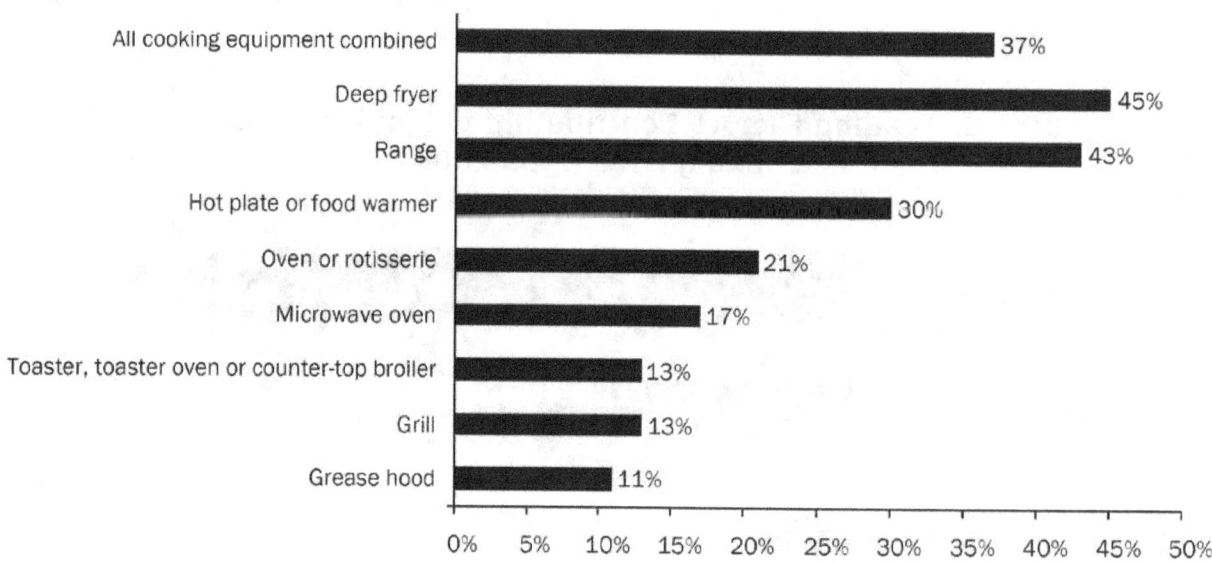

One-quarter of the fatal U.S. home cooking fire victims and one-fifth of the injured were in the area of origin and involved in the ignition.

Figure 17 shows that more than half (53 percent) of those killed and 41 percent of those injured in U.S. home cooking fires were described as not in the area of origin but involved in the ignition. This could be a description of unattended cooking. Twenty-four percent of the fatalities and 21 percent of the injuries were both in the area of origin and involved in the ignition, possibly injured in a fire that started while they were doing the cooking.

---

* The data used in causal analyses is based on nonconfined fires only.

Figure 17. Victim Location at Ignition in Home Cooking Equipment Fires: 1999-2003

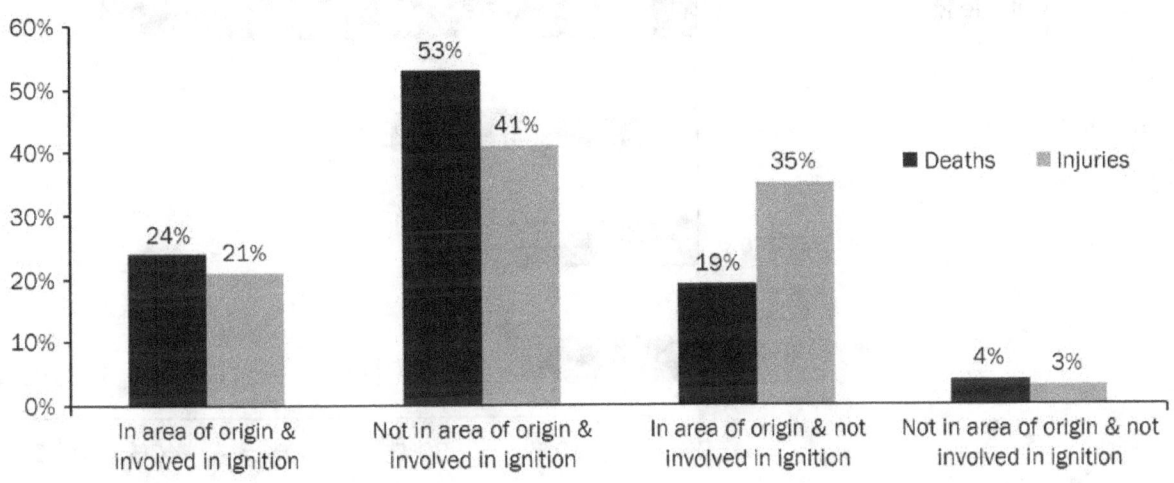

Figure 18. Home Cooking Equipment Fire Victims by Location at Time of Injury: 1999-2003

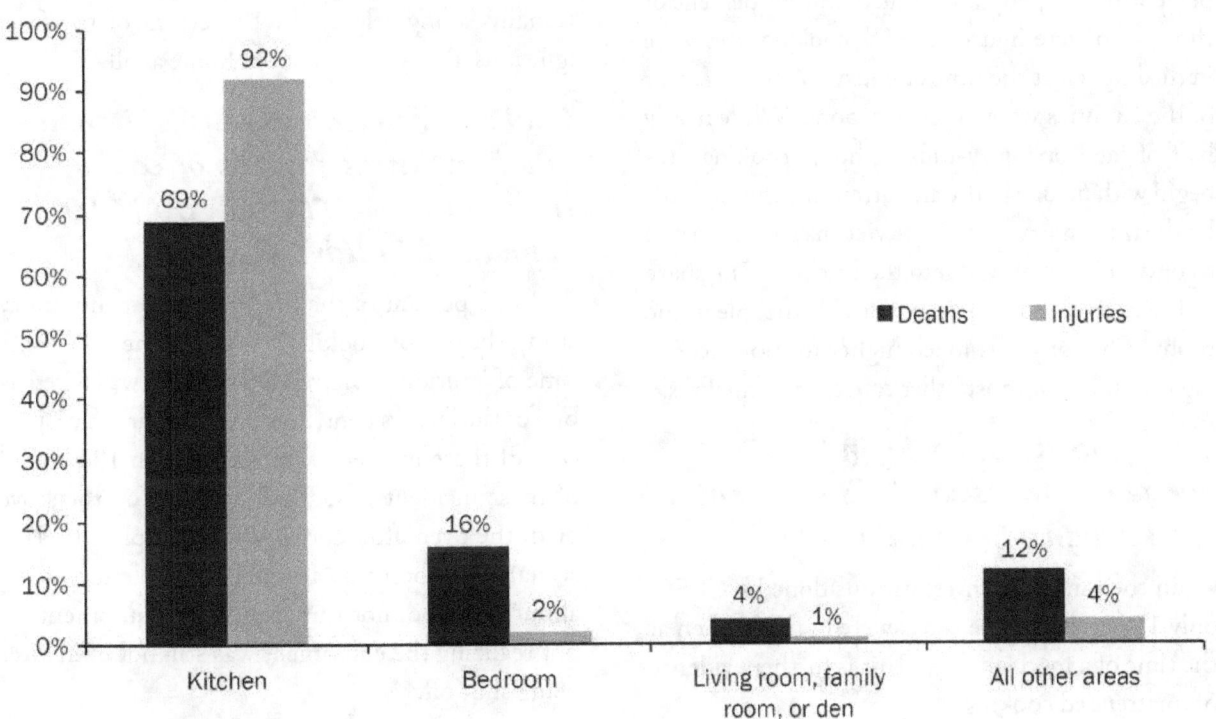

Figure 19. Location of Cook at Time of Food Ignition: CPSC Range Fire Study

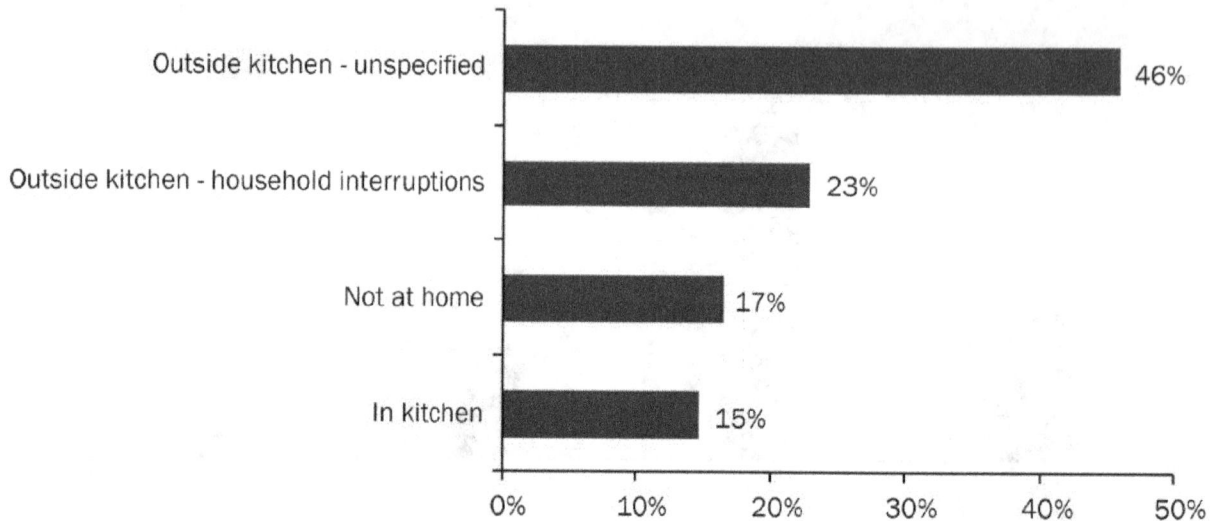

The vast majority of people injured in U.S. home cooking fires were in the kitchen when hurt.

Figure 18 shows that, from 1999 to 2003, 69 percent of the people who died and 92 percent of those who were injured in U.S. cooking fires were in the kitchen at the time of injury. Sixteen percent of the fatalities were in the bedroom. While nearly half of fatal or injury-causing home cooking fires begin with no one in the area attending the cooking, by the time injury occurs, the victims have returned to or otherwise moved into the kitchen. The share of fire deaths and injuries suffered by people in the room of fire origin is much higher for home cooking fires than for most other causes of home fires.

The majority of cooks in CPSC's range fire investigations were not in the kitchen when food ignited.

In the range fire investigations done by CPSC, only 15 percent of the cooks were in the kitchen at the time of a food ignition. This is further evidence of unattended cooking.

Figure 19 shows that almost half (46 percent) were outside of the kitchen at an unspecified location. In 23 percent of the food ignitions, the cook was outside of the kitchen due to household interruptions such as children, the phone, or unintentionally falling asleep. In 17 percent of the food ignitions, the cook was not at home at all.[8]

Cooks were also out of the kitchen in almost three-fourths of cooking fires in the NASFM and AHAM 10-community study.

In 73 percent of the fires in the 10-community study, the person cooking was not in the area at the time of ignition. Unattended cooking was listed as one of the factors contributing to the fire in 63 percent of these incidents. In addition, in 15 percent of these incidents, food had been left on the stove or in the oven after cooking was completed. This scenario can occur as a result of people forgetting about the food, not turning off the equipment, or not realizing the equipment was still hot even after being shut off.[7]

Cooks were distracted or forgot they were cooking in most Bay-Waikato, New Zealand, unattended cooking fires.

Seventy households in the Bay-Waikato Region of New Zealand who had experienced a reported kitchen fire participated in a study released in 1998. This study found that 51, or 73 percent, of these households said that their kitchen fire resulted from cooking. In 86 percent of these cooking fires, the cooking was unattended at the time of ignition. In 75 percent of the unattended cooking fires, the cooks forgot they were cooking or were distracted. In another 18 percent, the cook consciously left the cooking unattended.[20]

Separating Combustibles from Cooking Heat Sources

All home cooking fires involve a lack of sufficient control and a lack of sufficient separation. As discussed, some behavioral factors contributing to ignition emphasize the behavior or oversight that failed to keep cooking equipment properly controlled (e.g., unattended cooking, unintentionally turning on or not turning off equipment). Other behavioral factors contributing to ignition, however, emphasize the failure to keep combustibles separate from cooking heat sources (e.g., combustible too close to heat source).

As noted earlier, some type of combustible material too close to the cooking equipment was a factor in 13 percent of home cooking fires, 24 percent of the associated deaths, and 12 percent of the associated injuries, making heat source too close to combustibles the second leading factor contributing to ignition for home cooking fires, after unattended equipment.

When considering the issue of separation, it is useful to consider the types of items typically first ignited in home cooking fires.

Nearly half of all home range fires and half of associated injuries involve ignition of cooking materials and three behavioral factors contributing to ignition.

Unattended cooking equipment was a factor in 43 percent of range fires, unintentionally turned on or not turned off in 11 percent, and combustibles too close to heat source in 11 percent. Cooking materials were the items first ignited in 80 percent of the range fires caused by unattended cooking, 45 percent of the range fires in which the equipment was unintentionally turned on or on turned off, and 25 percent of the range fires caused by a heat source too close to combustible materials. Table 2 shows leading items first ignited for each of these three factors contributing to ignition.

Combustibles too close to heat can describe a variety of situations.

Although combustibles too close was a contributing factor in only 11 percent of the reported non-confined home cooking structure fires from 1999 to 2003, these fires caused 21 percent of the associated deaths. From 1999 to 2002, ranges or cooktops were involved in an average of 70 reported worn-clothing ignitions in home structure fires per year, resulting in an annual average of 36 deaths, 30 injuries, and $0.2 million in direct property damage. Although reported worn-clothing ignitions by stoves are unusual, on average, half resulted in a fatality. Overall, ranges or cooktops were involved in 14 percent of the ignitions of worn clothing, 30 percent of the associated deaths, 20 percent of the associated injuries, and 4 percent of the associated property damage.[21]

Older adults were the most common victims of clothing ignitions while cooking.

From 1999 to 2003, three-quarters of the people killed by the ignition of their clothing by cooking equipment were 65 or older, as were more than one-third of those who were nonfatally injured by that type of scenario. Less serious scenarios resulting from combustibles too close to the heat source involved normal kitchen supplies or clutter such as potholders, rags, trash, and containers that were ignited by the cooking equipment.

Nine percent of U.S. oven fires began with household utensils.

In 9 percent of both gas and electric oven fires reported from 1999 to 2003, household utensils, including kitchen and cleaning utensils, were the item first ignited. Six percent of the gas range fires and 5 percent of the electric range fires began with these utensils. The factor "improper container or storage" contributed to 5 percent of the gas oven fires and 3 percent of the electric oven fires, but only 1 percent of the gas range fires. This factor was not frequent enough to be listed with any other type of cooking equipment.[2]

Table 2. 1999-2003 U.S. Home Structure Fires Involving the Range and Selected Factors Contributing to Ignition, Excluding Confined Fires, by Item First Ignited Percents for Each Factor Contributing to Ignition and for Total Range Fires

Factor Contributing to Ignition	Item First Ignited	Percent Fires of Factor	of Total	Percent Deaths of Factor	of Total	Percent Injuries of Factor	of Total
Unattended cooking equipment	All items	100%	43%	100%	48%	100%	50%
	Cooking materials	80%	34%	48%	23%	90%	46%
	Cabinetry	4%	2%	11%	5%	1%	1%
	Household utensil	3%	1%	0%	0%	2%	1%
	Flammable or combustible gas or liquid	3%	1%	0%	0%	4%	2%
	Appliance housing	2%	1%	5%	2%	0%	0%
	Interior wall covering	2%	1%	11%	5%	0%	0%
	Unclassified item	2%	1%	0%	0%	1%	0%
Unintentionally turned on or not turned off	All items	100%	11%	100%	6%	100%	8%
	Cooking materials	45%	5%	0%	0%	57%	5%
	Household utensil	12%	1%	0%	0%	6%	1%
	Box or bag	9%	1%	0%	0%	9%	1%
	Appliance housing	7%	1%	0%	0%	0%	0%
	Cabinetry	5%	1%	50%	3%	0%	0%
	Unclassified item	4%	0%	0%	0%	7%	1%
	Interior wall covering	4%	0%	0%	0%	6%	0%
	Flammable or combustible gas or liquid	3%	0%	0%	0%	3%	0%
	Unclassified utensil or furniture	2%	0%	0%	0%	0%	0%

continued on next page

Factor Contributing to Ignition	Item First Ignited	Percent Fires of Factor	of Total	Percent Deaths of Factor	of Total	Percent Injuries of Factor	of Total
Heat source too close to combustibles	All items	100%	11%	100%	21%	100%	12%
	Cooking materials	25%	3%	0%	0%	18%	2%
	Box or bag	11%	1%	0%	0%	4%	0%
	Household utensil	10%	1%	0%	0%	5%	1%
	Unclassified item	7%	1%	0%	0%	8%	1%
	Papers	4%	0%	0%	0%	0%	0%
	Flammable or combustible gas or liquid	4%	0%	0%	0%	10%	1%
	Cabinetry	4%	0%	0%	0%	0%	0%
	Linen other than bedding	4%	0%	0%	0%	8%	1%
	Unclassified soft goods or clothing	4%	0%	13%	3%	9%	1%
	Clothing	4%	0%	39%	8%	15%	2%
	Appliance housing	4%	0%	0%	0%	5%	1%
	Interior wall covering	3%	0%	30%	6%	4%	0%
	Curtain or drapery	2%	0%	0%	0%	4%	1%
	Trash or waste	2%	0%	0%	0%	0%	0%
	Multiple items first ignited	2%	0%	0%	0%	4%	0%
	Unclassified utensil or furniture	1%	0%	0%	0%	0%	0%

Source: NFIRS, NFPA survey analyzed for this report

Use of Cooking Equipment for Other than Intended Purpose

Roughly one-fourth of people receiving energy assistance have used a kitchen stove for heat in the previous year.

Although NFIRS does not currently capture information on fires caused by cooking equipment being used for heat, there have been a number of these types of fires that have caused single or multiple fatalities. In fiscal year 2005, more than 4.9 million low-income households received financial assistance with heating and cooling bills through the Low Income Home Energy Association Program (LIHEAP). A survey of 1,100 LIHEAP recipients found that roughly one-quarter used a kitchen stove or oven to provide heat in at least one month in the past year because of a lack of funds for the energy bill.[22] This translates to roughly 1.2 million of the households who receive this financial assistance using a kitchen stove for heat in at least one month a year.

The frequency of using a kitchen stove for heat varied by region. Figure 20 shows that 34 percent of the respondents in the South reported using a stove or oven for heat in at least one month of the year compared to 26 percent in the West, 22 percent in the Northeast, and 18 percent in the Midwest.

Figure 20. Energy Assistance Recipient Use of Stove or Oven for Heat, by Region

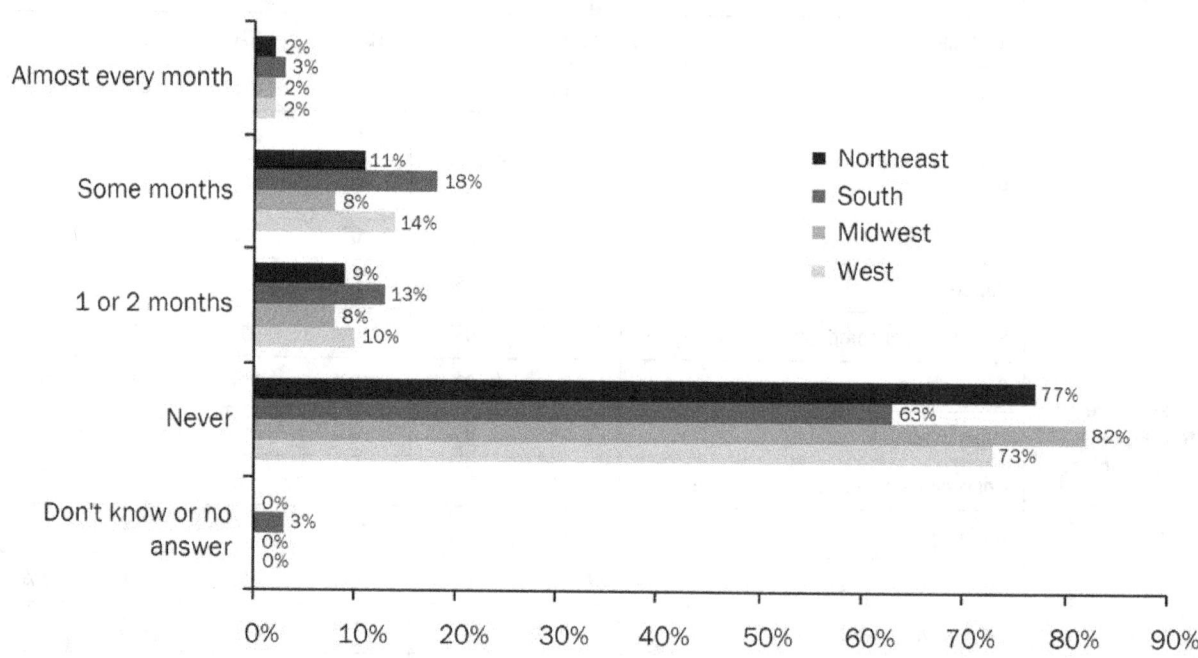

Types of Cooking Associated with Cooking Fires

Almost two-thirds of the food ignitions in CPSC's range fire study began with frying.

Combination units of cooktop and ovens were involved in 93 percent of CPSC's 289 range fire investigations, surface-only cooktops were involved in 5 percent, and separate oven units were involved in 2 percent. Overall, 218 (75 percent) of the range fires in this CPSC study began with food ignitions. Figure 21 shows that 63 percent of the range fire food ignitions occurred when someone was frying food. An additional 18 percent of the fires resulted from boiling and 10 percent resulted from baking. The oven was usually involved in baking fires while frying and boiling were usually done on the stovetop.[8]

Figure 21. Range Process in Food Ignitions: CPSC Range Fire Study

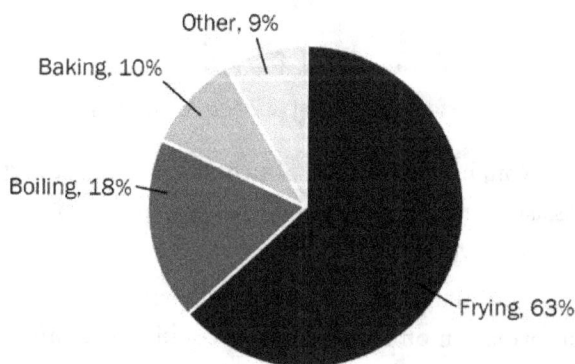

Cooking oil was the food ignited in almost half (46 percent) of cooktop food ignitions.

Figure 22 shows the type of food first ignited in the 192 cooktop food ignitions and 26 oven food

Figure 22. Foods Ignited in CPSC Range Fire Study by Type of Cooking Equipment

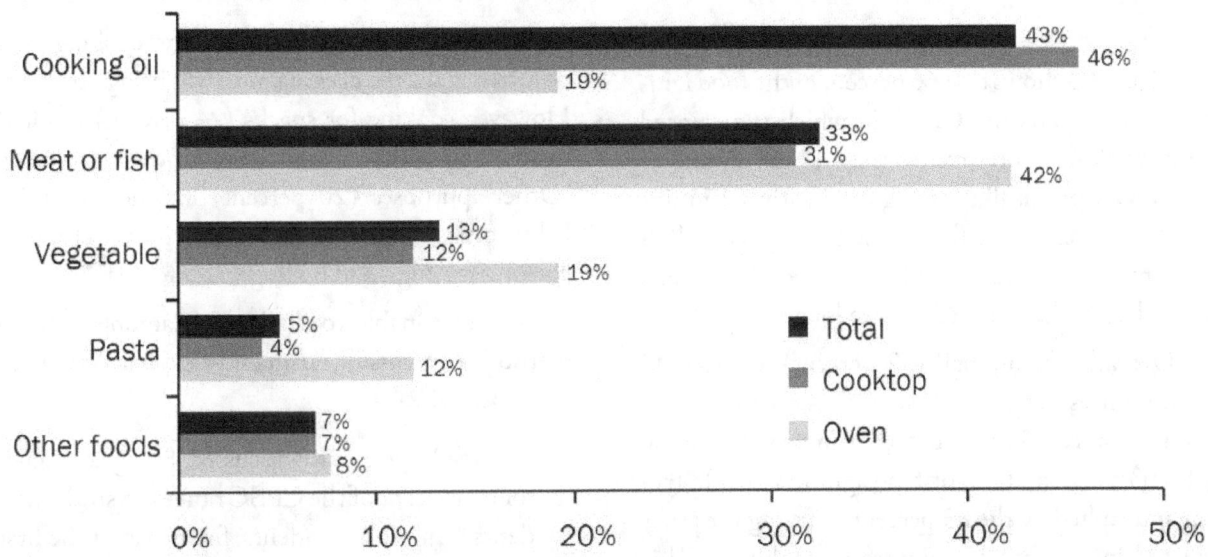

ignitions in the CPSC range fire study. Three-quarters of the food ignitions involved cooking oil, meat, or fish. [8]

Although over 40 percent of the cooking oil fires began when food was simmering in oil, the authors noted cooking oil fires frequently started before other food was added to the heated oil.

Bay-Waikato study identified different cooking oil and fat fire scenarios.

In 64 percent of the Bay-Waikato kitchen fires, cooks were either shallow frying (35 percent) or deep frying (29 percent). [20] More information on these Bay-Waikato oil or fat cooking fires follows:

+ In 30 percent of the fires, the cooks said they had oil or fat on the stove and forgot to turn off the heat.

+ Distractions from children, animals, phone calls, visitors, clean-up activities, etc., resulted in the oil or fat cooking unattended in 23 percent of the fires.

+ Twenty percent reported that the stove or element was on too high a setting.

+ Ten percent said the fire started when they were adding or removing food to or from the pan of oil or fat.

+ The burner or stove was unintentionally turned on in 10 percent of the fires.

+ In 7 percent of the fires, fat or grease had built up under the element and ignited when the stove was used again later.

Sixteen percent of the cooking fires in this study involved boiling. In some cases, those boiling had left the pan cooking on higher heat than intended, some forgot they were cooking, some intentionally left it unattended, and some thought they had turned off the heat when they had, in fact, turned it back on. Fires resulting from boiling started after the liquid had evaporated.

Ovens were involved in 10 percent of the cooking fires. In some cases, baking products fell or dripped onto the heating element. In other cases, nonfood items had been left in the oven.

The majority of food ignitions from frying or baking in the CPSC study occurred in the first 15 minutes of cooking.

Figure 23 shows that 66 percent of the food ignitions investigated by CPSC in which the elapsed time was known occurred in the first 15 minutes of cooking. Specifically, food ignited within 15 minutes in 83 percent of fires caused by frying and 88 percent of fires caused by baking. Food ignitions caused by boiling tended to take longer.[8]

Overall, roughly half (52 percent) of the food ignition fires in this study were fires caused by frying that started in the first 15 minutes of cooking (the 63 percent share of food ignitions while frying multiplied by the 83 percent of frying fires that started in the first 15 minutes of cooking). This shows how important it is to pay close attention when frying.

Twenty-four percent of the Bay-Waikato, New Zealand, cooking fires occurred during snack preparation.

Almost half (49 percent) of the cooking fires occurred while the evening meal was being prepared. However, cooking for snacks (24 percent) resulted in more fires than cooking for lunch (8 percent). "Other" purposes (20 percent) include sterilizing baby bottles, heating water or fat as part of cleaning, cooking for dogs, and boiling water for tea.[20]

It is reasonable to distinguish among cooking methods in terms of estimated risk, with frying as the most risky.

♦ Frying. As previously discussed, frying accounted for 63 percent of the CPSC range fire study incidents.[8] In those incidents, fire began in the first 15 minutes for 83 percent of the fires, while 12 percent began at least 30 minutes after cooking

Figure 23. Elapsed Cooking Time before Food Ignition by Cooking Process in CPSC Range Fire Study

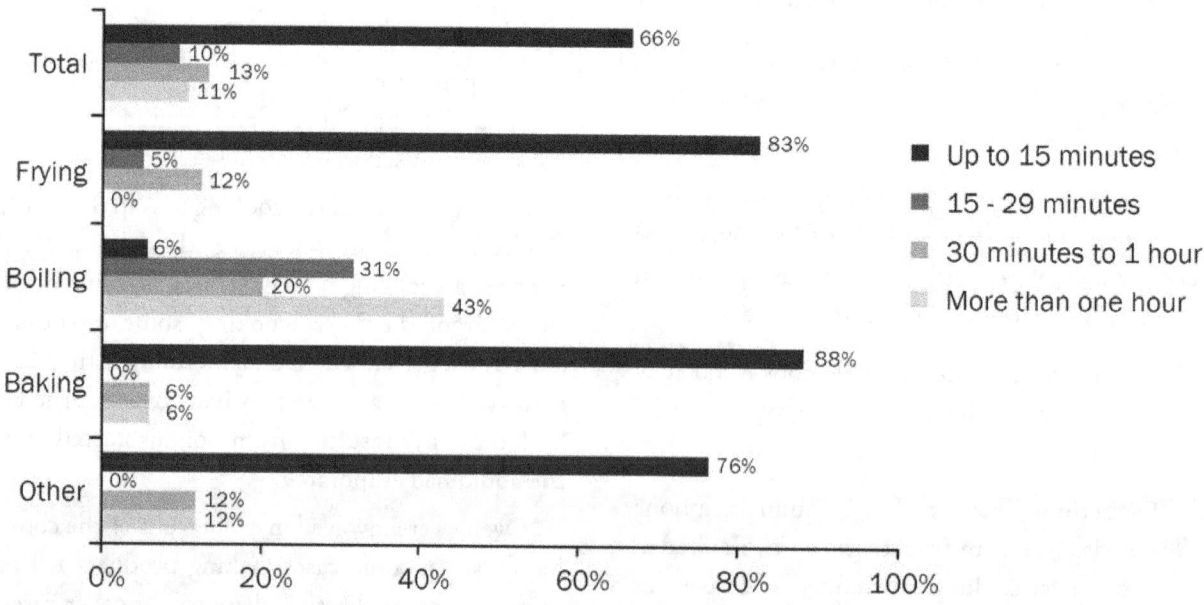

began. Frying inherently involves a combustible medium in addition to the food, namely the cooking oil, and two-thirds of the CPSC range fire frying incidents began with ignition of the cooking oil. In addition, a frying pan provides no containment for fire if one begins. For these reasons, there can be no exceptions to attendance at frying by the cook. Because frying is relatively quick, there should be no great hardship in attendance.

Deep fryers involve larger quantities of hot cooking oil than that involved in regular frying, and turkey fryers involve extremely large quantities of hot cooking oil. Because the frying process involves inserting the food into the heated medium, then later removing it and transferring it to a drying location, deep frying with these larger quantities of hot oil involve numerous opportunities for thermal burns and scalds, as well as fire ignitions. As a result, while consumer use of these products is strongly discouraged, there also can be no exceptions to attendance if used.

Woks and other devices designed for stir-fry cooking also need to be considered within the frying cooking method and need to be attended closely.

- **Broiling and grilling.** Broiling and grilling were part of the "other" category that accounted for 9 percent of the CPSC range fire study incidents.[8] (The dictionary defines grilling as "broiling on a gridiron."[23]) In the "other" incidents, fire began in the first 15 minutes for 76 percent of the fires, while 24 percent began at least 30 minutes after cooking began. Broiling and grilling do not inherently involve a combustible medium in addition to the food. However, both types of cooking often involve a need for regular cook intervention, such as turning the food, in order to avoid overheating. Broiling is sometimes done in an oven, which provides some containment for fire if one begins. However,

when broiling in an electric oven, the oven door is left ajar, limiting the containment. In addition, other broiling and all grilling are done on exposed cooking surfaces. For all these reasons, broiling and grilling can be regarded as only slightly less risky than frying, and there should be no exceptions to attendance.

Barbecue grills are designed for use outside, and that location may reduce the risk, if fire occurs, of fire spread from the grill to other combustibles. In addition, fatal barbecue grill fires are rare. However, when grill fires do occur, they nearly always involve ignition of a part of a structure. Indoor use of charcoal grills, specifically, also introduces a significant risk of death due to carbon monoxide buildup. This combination accounts for more than 10 deaths a year.

- **Baking and roasting.** Baking accounted for 10 percent of the CPSC range fire study incidents.[8] (Baking and roasting are both defined as "cooking with dry heat."[23] This presumably refers to convective heat, as contrasted with the radiant heat used in broiling and grilling.) Fire began in the first 15 minutes for 88 percent of the fires, while 12 percent ignited at least 30 minutes after cooking began. Normally baking and roasting do not inherently involve a combustible medium in addition to the food. Baking does not involve a need for regular cook intervention, but some roasting does require regular cook intervention, such as basting, in order to avoid overheating. Baking and roasting typically are done in an oven, which provides containment for fire if one begins. Primarily for this last reason, baking and roasting can be regarded as less risky than broiling and grilling. Brief absences during cooking, which tends to take longer than frying, broiling, or grilling, can be justified, provided a timer is used to remind the cook to check on the cooking.

Toaster ovens can be regarded as small baking devices, although they can be used for broiling as well. Hot plates and food warmers involve conducted heat rather than convective heat. Together with toasters and toaster ovens, they account for most of the fires and related deaths associated with portable cooking or warming devices. Hot plates and toasters should not be left unattended during their typically very short cooking periods.

+ Boiling. Boiling accounted for 18 percent of the CPSC range fire study incidents.[8] Fire began during the first 15 minutes in 6 percent of the fires, while 63 percent ignited at least 30 minutes after cooking began. Boiling does not inherently involve a combustible medium in addition to the food. In fact, the normal medium of water typically will prevent fire until or unless it boils away. Normally, boiling does not involve a need for regular cook intervention. Boiling may be done either in an enclosed container (e.g., kettle, coffeemaker) or in an open container (e.g., pan). However, if the water boils away, the container may fail and deform, removing the containment. Primarily because few fires occur early in the boiling process, boiling can be treated as comparable to or less risky than baking and roasting. Brief absences during cooking can be justified, provided a timer is used to remind the cook to check on the cooking. Unlike other types of cooking, the periodic inspection can identify an impending hazard easily (i.e., the imminent loss of the water), with ample time to correct the problem.

Simmering is cooking done at or just below the boiling point. If the simmering temperature is well below the boiling point, simmering is like slow cooking (see below) or even food warming. "Stewing" is defined as slow boiling. "Steaming" is cooking by exposure to steam, i.e., water in the form of heated vapor. Each of these presents a variation on boiling.

+ Slow cooking. Slow cooking was not identified in the CPSC range study and represents a small share of the estimated home fires involving all types of portable cooking or warming equipment. Heat levels typically are low enough so that other provisions for safety, including close attendance, are not necessary. If the cookware is placed where an unlikely minor overflow will not contact other combustibles, there will be added safety. If a crock pot or similar device is used, any ignition of food also will be contained, provided nothing has interfered with the equipment itself.

Alcohol and Other "Human Factors" Associated with Cooking Fires

Alcohol or other drugs were mentioned as possible* human factors contributing to ignition in fires, causing 20 percent of the U.S. home cooking fire deaths.

Although impairment by alcohol or drugs was noted as a possible factor contributing to ignition in only 2 percent of the fires, Figure 24 shows that such a condition was listed as a possible contributing factor in 20 percent of the associated deaths and 6 percent of the associated injuries.

Falling asleep was a factor in 6 percent of the cooking fires and 27 percent of the associated deaths. Falling asleep and impairment by drugs or alcohol can result in cooking being left unattended. Physical disability was a factor contributing to injury in fires resulting in 10 percent of the deaths.

---

* Some of the human factors (impairment, mental disability) are defined with the qualifier "possibly" in the name of the code, while others are not.

Figure 24. Human Factors Associated with Cooking Equipment Fires: 1999-2003

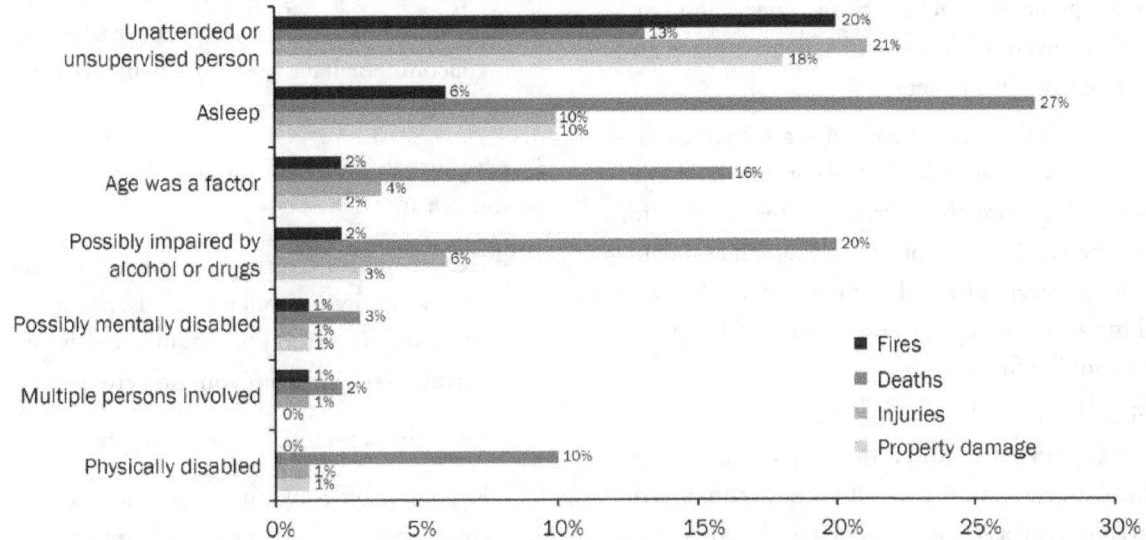

An unattended or unsupervised person was said to have contributed to the ignition in 20 percent of these fires. Given that age was a factor contributing to ignition in only 2 percent of the home cooking fires, it is possible that some or most of these "unattended or unsupervised person" factors actually refer to unattended cooking. Unfortunately, however, this cannot be confirmed with available data. In the 10-community study, possible drug or alcohol was mentioned as a factor in 6 percent of the fires.[7] The authors note that this condition is historically underreported, as definitive evidence such as blood alcohol levels are not generally available to fire officials.

Summary Discussion

Unattended cooking is the leading factor contributing to cooking fires and can arise in a number of ways, large and small—not being home at all, forgetting that cooking is still going on, being distracted by household interruptions, mistakenly believing cooking has been turned down or off when it has not been, and deliberately choosing to leave cooking unattended, presumably because of a lack of appreciation of the risks involved.

The failure to keep combustibles from a heat source also plays a significant role in the cooking fire problem. If a heat source is not sufficiently close or hot enough to bring a combustible item to its ignition temperature, no fire will occur. Therefore, sufficient separation between combustibles and cooking heat sources should be encouraged.

Although it is safest to pay constant attention to all cooking, the dangers of unattended cooking vary somewhat in degree by type of cooking method. Frying, the most common type of cooking cited when cooking fires occur, involves a combustible medium in addition to the food, and no containment in the cooking vessel if fire occurs. As a result, there can be no exceptions to attendance at frying by the cook. Broiling and grilling usually require frequent interaction by the cook (e.g., rotating the meat) to keep heating even and avoid overheating in any one area. This means a risk of overheating is to be expected if cooking is not closely attended. Simmering, baking, roasting, and boiling do not involve an additional combustible medium and often involve lower temperatures and/or a cooking vessel designed for containment and for extended periods without supervision. They still require supervision but not necessarily as continuously. In addition, some types of cooking equipment may be more forgiving of a lack of close supervision (e.g., microwave oven, which provides containment and shuts itself off on a timer).

Often, alcohol and other drugs are cited as factors in cooking fires or in the fatal or non-fatal injuries resulting from these fires. Other reasons for diminished ability to control cooking safely, including falling asleep, physical or mental disability, and the limitations of age, also are cited as factors in home cooking fires.

Finally, although statistics are not available on how many fires are caused by people using stoves for heat, an estimated 1.2 million households who received energy assistance had used a kitchen stove for heat in at least one of the previous 12 months.

Behavioral Strategies

The following specific messages arising from this chapter address preventing unattended cooking, preventing cooking by people with less than full capacity to supervise, keeping things that can catch fire away from heat sources, what you should do if your clothes catch fire, and preventing usage of equipment for unintended purposes:

1. Watch what you heat!

   + The leading cause of fires in the kitchen is unattended cooking.

   + Stay in the kitchen when you are frying, grilling, or broiling food. If you leave the kitchen for even a short period of time, turn off the stove.

   + If you are simmering, baking, roasting, or boiling food, check it regularly, remain in the home while food is cooking, and use a timer to remind you that you're cooking.

2. Stay alert.

   To prevent cooking fires, you have to be alert. You won't be if you are sleepy, have been drinking alcohol, or have taken medicine that makes you drowsy.

3. Keeping things that can catch fire and heat sources apart.

   + Keep anything that can catch fire—potholders, oven mitts, wooden utensils, paper or plastic bags, food packaging, towels, or curtains—away from your stovetop.

   + Keep the stovetop, burners, and oven clean.

   + Keep pets off cooking surfaces and nearby countertops to prevent them from knocking things onto the burner.

   + Wear short, close-fitting or tightly rolled sleeves when cooking. Loose clothing can dangle onto stove burners and catch fire if it comes into contact with a gas flame or electric burner.

4. What to do if your clothes catch fire.

   If your clothes catch fire, stop, drop, and roll. Stop immediately, drop to the ground, and cover face with hands. Roll over and over or back and forth to put out the fire. Immediately cool the burn with cool water for 3 to 5 minutes and then seek emergency medical care.

5. Use equipment for intended purposes only.

   Cook only with equipment designed and intended for cooking, and heat your home only with equipment designed and intended for heating. There is additional danger of fire, injury, or death if equipment is used for a purpose for which it was not intended.

Chapter 5

Civilian Firefighting and Fire Extinguishment

ome cooking fires are more likely than fires of any other cause to lead to injuries because of occupant attempts to control or extinguish the fire themselves.

More than half of the reported U.S. home cooking fire injuries occurred when individuals tried to fight the fire themselves.

Figure 25 shows that, from 1999 to 2003, 55 percent of the people injured in reported U.S. home cooking equipment structure fires were injured while trying to fight the fire themselves. This is five times

the 11 percent injured while trying to fight the fire in home fires not caused by cooking. Twelve percent of the nonfatal cooking fire injuries occurred while escaping. However, this is only one-third of the percent injured while escaping noncooking fires.

Figure 25 also shows that 11 percent of the people who died as a result of home cooking fires were fatally injured while attempting to fight the fire themselves, compared to only 1 percent of the noncooking fire fatalities. The 14 percent of cooking fire fatalities who were trying to escape is less than half the 33 percent of the noncooking fire fatalities who tried to flee. Forty-one percent

Figure 25. U.S. Home Cooking Fire Victims by Leading Activity at Time of Injury: 1999-2003

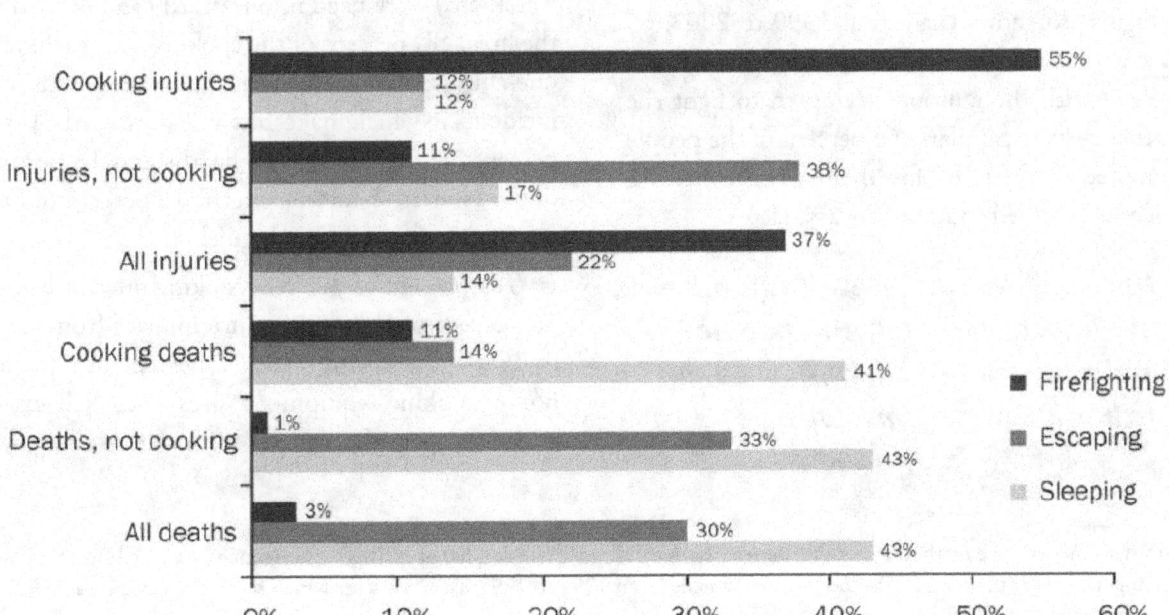

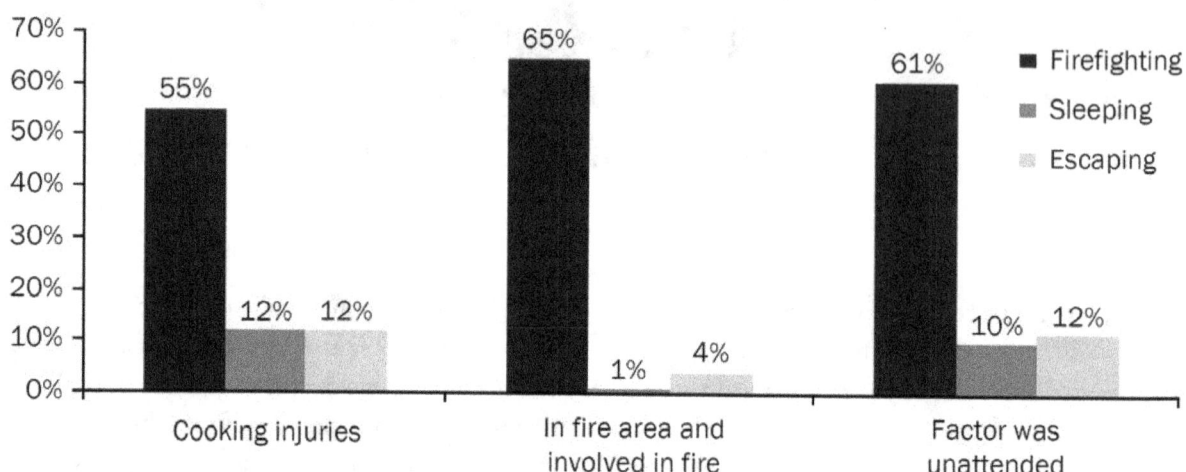

Figure 26. Home Cooking Fire Injuries by Victim Location and Activity at Time of Injury: 1999-2003

of people killed in home cooking fires had been asleep. However, little difference is seen between this and the 43 percent of people killed while sleeping in noncooking fires.

Individuals were almost as likely to try to fight a cooking fire in unattended cooking fires as they were when they had been in the fire area.

Figure 26 shows that, from 1999 to 2003, 65 percent of those who were in the fire area and involved with the ignition attempted to fight the fire themselves. Similarly, 61 percent of the people attempted to fight the fire themselves in fires in which unattended equipment was a factor.

In the Bay-Waikato, New Zealand, study, more than half the people who discovered kitchen fires tried to put the fire out themselves.

In more than half of the Bay-Waikato kitchen fires, the person discovering the fire attempted to extinguish the fire. Only 18 percent waited for the fire service.[20]

Makeshift aids or extinguishers put out half the reported U.S. home cooking fires.*

Figure 27 shows that, from 1994 to 1998, makeshift aids, which might include lids, baking soda, water, etc., were used in one-third (33 percent) of the fires. Six percent of the cooking fire deaths and one-third (33 percent) of the injuries resulted from incidents in which makeshift aids were used. Three percent of the cooking fire deaths and 18 percent of the injuries resulted from the 18 percent of fires put out by a portable extinguisher.

Ten percent of the U.S. cooking fire deaths and 16 percent of the cooking fire injuries from 1994 to 1998 resulted from the 27 percent of reported home cooking equipment fires that self-extinguished or, in other words, went out on their own

---

* Information about the method of extinguishment is not collected in Version 5.0 of NFIRS. Consequently, an analysis of method of extinguishment of reported U.S. cooking fires was done on 1994 to 1998 data, years for which this information was required in earlier versions of NFIRS.

Figure 27. Home Cooking Fires by Method of Extinguishment: 1994-1998

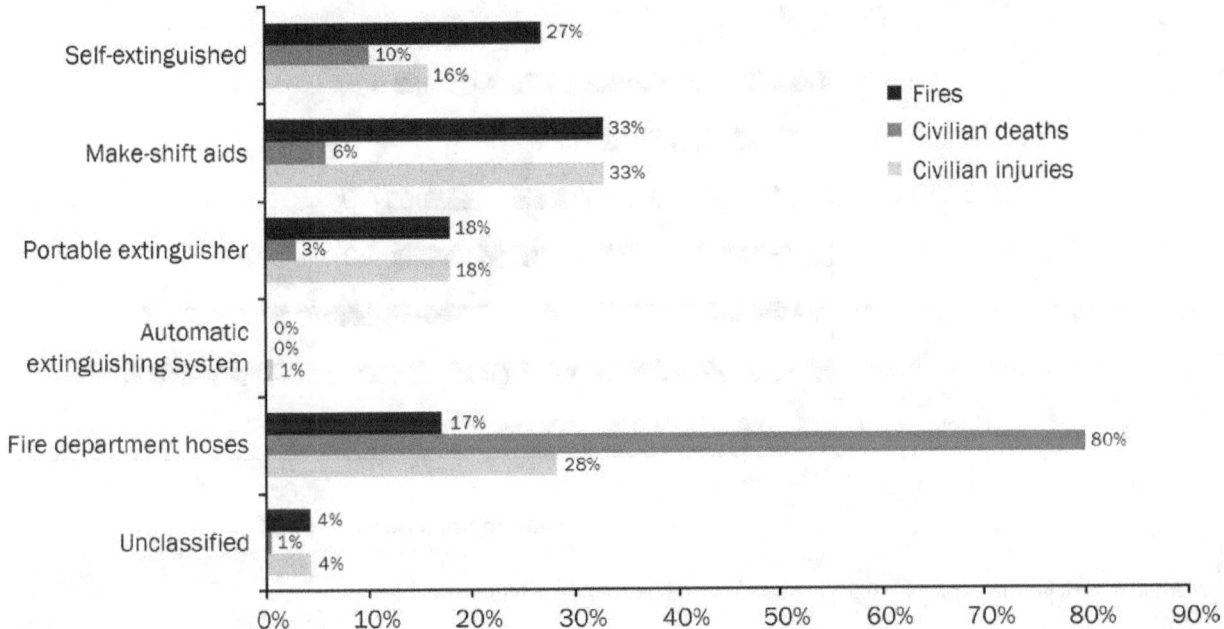

without any extinguishing agent being applied by a resident or firefighter.

Only 17 percent of the cooking fires were extinguished with fire department hoses, but these fires resulted in 80 percent of the associated deaths and 28 percent of the injuries. This is not surprising because a fire requiring use of a hose will tend to be a larger fire, with more potential to kill or injure occupants before the fire department arrives.

The patterns in Figure 27 are consistent with the idea that many people use extinguishers or makeshift aids successfully to control fires early and keep them small. At the same time, however, many injuries are occurring during these resident attempts to control fires themselves.

There are not enough data to tell how many injuries occur because residents tried to fight a fire after it had grown too large for resident control, because residents used flawed methods in their firefighting, or because even successful, appropriate firefighting

using best methods resulted in some kind of injury. However, these data do suggest that it would be an over-reaction to try to discourage residents from firefighting in all circumstances. So, while it is always safest to get away from a fire and outside of a burning structure, it would be appropriate to devote some educational resources to teaching people how to fight fires safely and effectively.

Makeshift aids and portable extinguishers had identical injury rates per 100 fires.

Figure 28 shows that the overall civilian injury rate for reported cooking fires was 4.8 per 100 fires from 1994 to 1998. Not surprisingly, the rate was lowest for self-extinguished fires. Fires extinguished with makeshift aids and portable extinguishers both had 4.9 injuries per 100 fires. And, again, the injury rate was highest for fires extinguished with fire department hoses at 7.8 injuries per 100 fires.

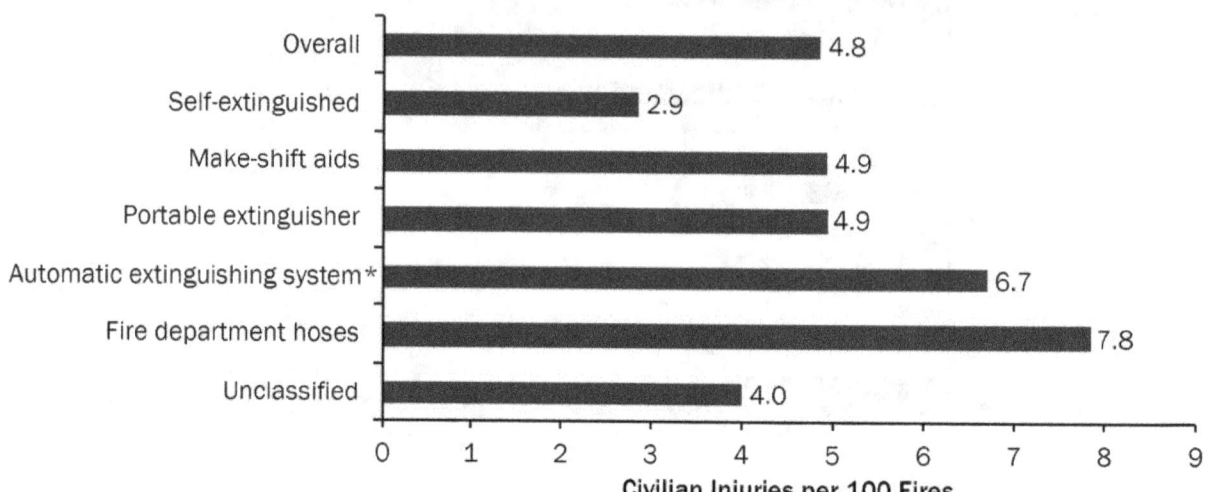

Figure 28. Home Cooking Fire Civilian Injury Rate by Method of Extinguishment: 1994-1998

*Method of extinguishment in only 0.5% of these fires and 0.6% of these injuries

The rate for automatic extinguishing systems seems high, but only 0.5 percent of the fires were extinguished by this equipment, and fires must reach a certain minimum size before automatic extinguishing equipment will operate. This means that, like those that have to be extinguished with department hoses, the average size of such fires may be greater—and the associated risk of injury may be greater—than with fires extinguished by portable extinguishers or makeshift aids.

Figure 29. Extinguishment Method Used in CPSC Study of Reported and Unreported Fires: December 1983-November 1984

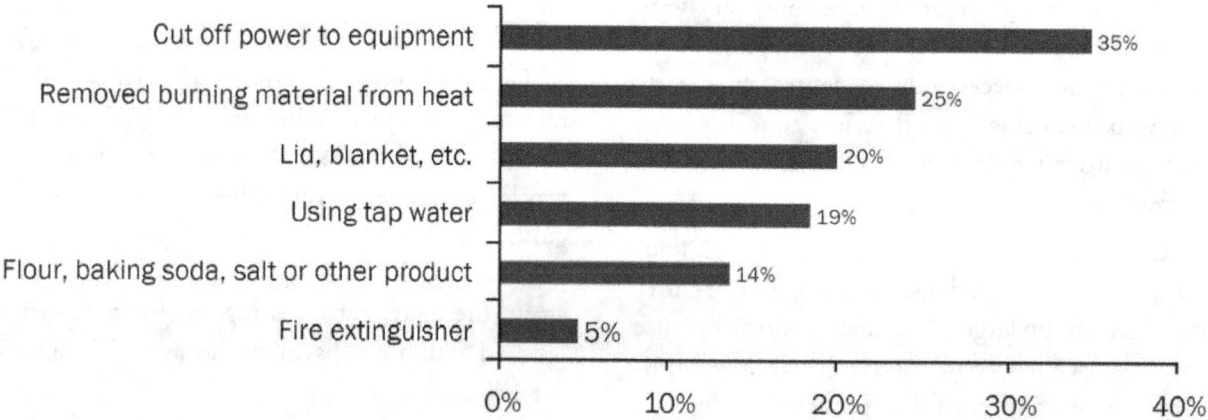

Eighty-seven percent of people with reported and unreported fires attempted to put the fire out themselves.

The study of unreported residential fires done for CPSC also compared reported and unreported fires. In 87 percent of the total (reported and unreported) residential fires, residents attempted to extinguish the fire themselves. They called the fire department in only 4.5 percent of the fires.[4]

Residents were asked about all of the approaches they had used to try to extinguish the fire. Figure 29 shows that 35 percent reported cutting or turning off power to the equipment; 25 percent reported removing the burning material from the heat source; 20 percent used a lid or blanket to smother the flames; 19 percent used tap water; 14 percent used baking soda, salt, flour, or some other product; and only 5 percent used fire extinguishers.

Almost two-thirds of those in the NASFM and AHAM 10-community study of reported cooking fires did not fight the fire themselves.

In the study of reported cooking fires in 10 communities, 64 percent said that, after ignition,

they "did not fight fire/left area."[7] However, Figure 30 shows that, of the 36 percent who attempted to suppress the fire themselves, half extinguished the fire properly, 19 percent opened the door on an oven fire, 14 percent used water, 4 percent used flour, and 12 percent were said to have used another improper agent. Unfortunately, the survey question did not ask if the occupant "put a lid on it," which is the preferred approach to dealing with small stovetop fires. Consequently, it is unclear what methods were actually used for proper extinguishment.

Nineteen percent of Bay-Waikato, New Zealand, study respondents who fought their kitchen fires tried to move the burning item.

Fifty-three percent of the individuals in the Bay-Waikato, New Zealand, study who tried to extinguish the fire themselves reported taking one or more "appropriate actions," including smothering the fire with wet towels or blankets, lids, or dirt; turning off the appliance or the main power; leaving the building; and shutting doors.[20]

Forty-four percent reported actions that could be dangerous. This included 19 percent who tried to move the burning item. Others put inappropriate materials, usually water, but also salt, flour,

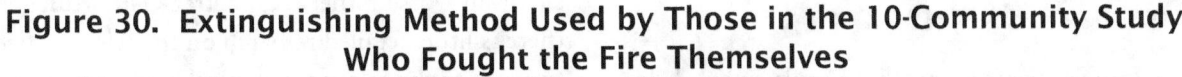

Figure 30. Extinguishing Method Used by Those in the 10-Community Study Who Fought the Fire Themselves

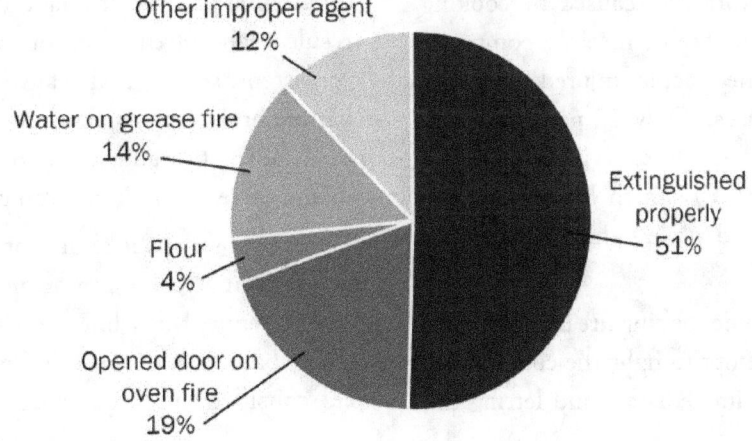

Figure 31. Automatic Suppression System Performance When Present in Home Cooking Fires: 1994-1998

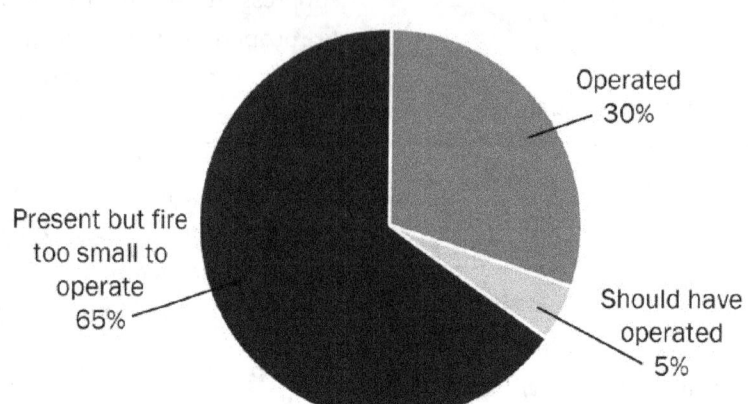

and baking powder, on oil fires. Some entered or returned to a burning building, and some took the lid off or opened the door on a burning item.

The fire was too small to activate sprinklers in two-thirds of U.S. home cooking fires with this equipment.

Figure 31 shows that when automatic extinguishing equipment was present, the equipment operated in 30 percent of the home cooking fires from 1994 to 1998. In almost two-thirds of the fires, the fire was too small to activate the equipment.

Summary Discussion

Fifty-five percent of the people injured in reported home structure fires caused by cooking were trying to fight the fire themselves, compared to 11 percent of the people injured in fighting fires of other causes. Only 12 percent of the people killed by these fires were trying to fight the fire themselves, but this is much larger than the 1 percent fatally injured while fighting home fires from other causes.

The majority of home cooking fire injuries occur when individuals attempt to fight the cooking fire themselves. Leaving immediately and letting the

fire department handle the fire greatly reduces the chance of civilian injury. However, resident firefighting plays a large role in successfully controlling or extinguishing fires while they are small. This supports the development of a behavioral strategy emphasizing safer resident firefighting in addition to one that aims to completely eliminate resident firefighting.

There are many messages, often contradictory, in circulation about the best way to handle kitchen fires. These messages can leave people unsure about how to proceed, or even lead to demonstrably unsafe firefighting practices that will make the situation worse rather than better. Unfortunately, there is little detailed research on the relative effectiveness or the relative injury risks associated with different approaches to handling small fires. As a result, many of the decisions required to develop consistent, sound, and realistic advice on how to handle and possibly fight cooking fires, must be made at the best judgments of experts rather than definitive research directly on point.

The consensus is clear that water should never be used on a grease fire or on fires with electrical components. But while some experts recommend using baking soda or salt on certain fires, others consider this impractical or even dangerous.

Smothering a fire with a lid seems to be an accepted approach. And, while the possibility of burns exists, a properly selected pan lid can cover the fire in one motion and can be used to shield the hand and arm of the resident while the lid is being put in place. In addition, fire blankets are routinely recommended in Europe and Australia but less often mentioned in the U.S.

Fire extinguishers also are recommended often but, when used incorrectly, they actually can spread a fire. It is important that individuals who would consider using a fire extinguisher in a fire situation receive training in how to use these devices properly. It also is important to ensure that this equipment is maintained properly and is operational. Many of the sources available mention fire extinguishers in passing, but most provide little specific guidance on how to use such equipment. While hands-on training is the best way to learn to use fire extinguishers, it is likely that many people who have these devices have not received any kind of training at all on their use.

Behavioral Strategies

The following specific messages arising from this chapter address how and when to fight cooking fires:

+ When in doubt, just get out. When you leave, close the door behind you to help contain the fire. Call 9-1-1 or the local emergency number after you leave.

+ If you do try to fight the fire, be sure others are already getting out and you have a clear path to the exit.

+ Always keep an oven mitt and a lid nearby when you are cooking. If a small grease fire starts in a pan, smother the flames by carefully sliding the lid over the pan (make sure you are wearing the oven mitt). Turn off the burner. Do not move the pan. To keep the fire from restarting, leave the lid on until the pan is completely cool.

+ In case of an oven fire, turn off the heat and keep the door closed to prevent flames from burning you or your clothing.

+ If you have a fire in your microwave oven, turn it off immediately and keep the door closed. Never open the door until the fire is completely out. Unplug the appliance if you can safely reach the outlet.

+ After a fire, both ovens and microwaves should be checked and/or serviced before being used again.

Additional educational messages related to civilian firefighting of cooking fires have been developed and publicized by a wide variety of national organizations, local fire departments, general safety organizations, burn prevention specialists, and popular media. Some of these messages are detailed in Appendix B.

Chapter 6

. . . .

Smoke Alarms and Fire Discovery

Smoke alarms play an important role in reducing deaths from cooking fires. One-third of the home cooking fire fatalities resulted from fires reported between 11 p.m. and 4 a.m. when many people are sleeping. Working smoke alarms could have prevented many of these deaths. Smoke alarms also can remind distracted individuals of food forgotten on the stove. Unfortunately, unwanted activations during cooking have caused too many people to disable their smoke alarms.

Smoke alarms were more likely to have operated in cooking fires than in other reported home fires in the U.S.

Figure 32 shows that home smoke or other fire alarms operated in 70 percent of the cooking fires reported to U.S. fire departments from 1999 to 2003. This may, however, underestimate the true performance and capability of smoke alarms. Confined cooking fires were coded as no working smoke alarms present if the smoke alarms did not alert the occupants, but confined cooking fires are especially likely to be discovered by occupants long before a working smoke alarm would be expected to activate. Also, some occupants may not have been home to be alerted. Even so, the 70 percent of smoke alarms that operated in cooking fires is considerably higher than the 43 percent found in 1999 to 2002 home fires of all causes.[24]

Operating smoke alarms were found in fires that caused only 23 percent of the home cooking fire deaths. However, 56 percent of the reported home cooking fire injuries resulted from home fires with working smoke alarms. In many cases, it appears that the sound of the smoke alarm alerted the occupants to fires that seemed small enough to try to handle without calling the fire department. As noted earlier, 55 percent of the nonfatal cooking injuries were incurred by civilians trying to fight the fire themselves.

Unwanted smoke alarm activations from cooking are a problem.

In a 2004 survey conducted for the NFPA, 40 percent of the respondents with smoke alarms reported that one had sounded at least once in the past 12 months. Sixty-nine percent reported activations due to routine cooking activities. Of the respondents who reported that an alarm had sounded, 24 percent thought that food had burned.[25] When smoke alarm batteries were missing, it was usually because of these annoying alarm activations from cooking. Researchers who visited households for a CPSC smoke alarm study found that the leading reason for battery removal or disconnect was unwanted activations. The leading problems cited for smoke alarms with missing batteries or disconnected power sources were 1) alarming to cooking fumes and 2) alarming continuously when powered.[26]

Figure 32. Home Cooking Fires by Smoke Alarm Status: 1999-2003

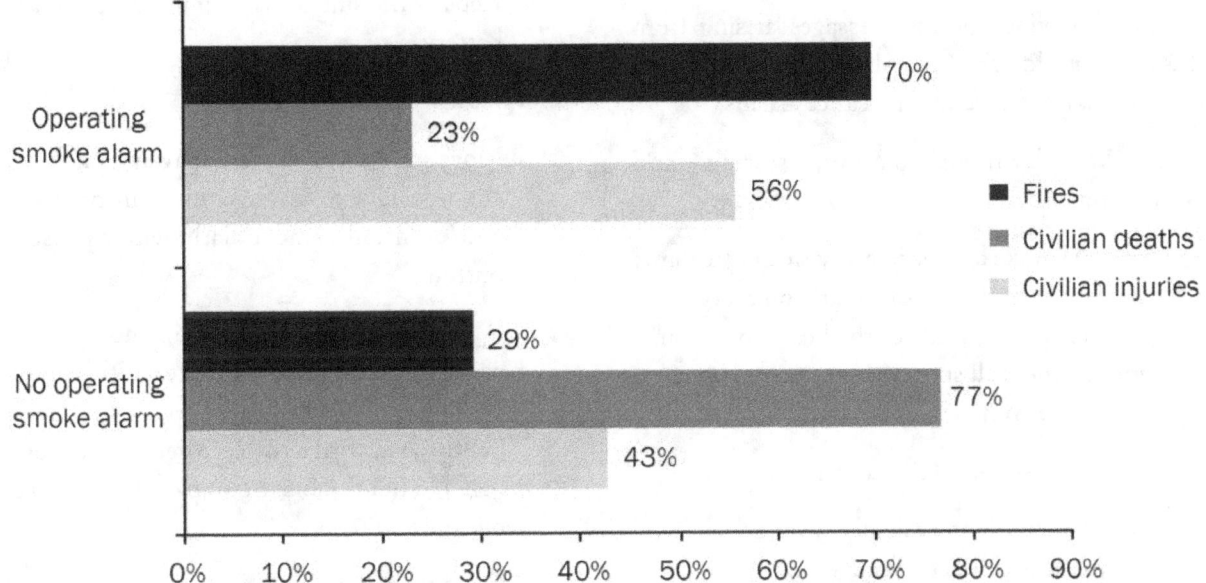

Summary Discussion

While working smoke alarms were present in 70 percent of the home cooking fires reported to U.S. fire departments from 1999 to 2003, they were present in only 23 percent of the home cooking fatalities. A working smoke alarm might have prevented many of the remaining 77 percent of fatalities.

The early warning of a smoke alarm may alert an individual to a fire that seems small enough to handle alone. Some of those fires will be controlled so quickly that they are never reported to the fire department while others may result in an injury to an occupant who might not have attempted to fight the fire if it had not been detected so early.

Some of the so-called nuisance activations, particularly from cooking, fall into a gray area. A sounding smoke alarm may remind a cook who has left the kitchen area of food on the stove requiring immediate attention. One-quarter of the people who had a smoke alarm go off in the past year reported their first thought was that food had burned. While not yet a fire, the potential exists if corrective action is not taken. If such action is taken, the situation often can be resolved quickly without fire department involvement.

However, unwanted smoke alarm activations during cooking too often result in disabled smoke alarms. In many nuisance activations, smoke alarms were found to be too close to a stove or bathroom. Moving the smoke alarm further away from these areas can reduce the number of these incidents. In addition, many smoke alarms have pause buttons that, when pressed, deactivate the smoke alarm for a few minutes. The smoke alarm then reactivates automatically. When the occupants know that the situation is not a real fire, pressing the pause button allows them to stop the noise without disabling the smoke alarm.

Behavioral Strategies

The following specific messages arising from this chapter address smoke alarm installation, testing, and maintenance, and nuisance alarms:

1. Smoke alarm installation, testing and maintenance.

    + Install smoke alarms in every sleeping room, outside each sleeping area, and on every level of your home. For the best protection, interconnect all smoke alarms throughout the home. When one sounds, they all sound.

    + Test each smoke alarm at least monthly.

    + Install a new battery in all conventional alarms at least once a year.

    + If the smoke alarm chirps, install a new battery in a conventional smoke alarm. Replace the smoke alarm if it has a 10-year battery.

2. Nuisance alarms.

    + Move smoke alarms farther away from kitchens according to manufacturers' instructions and/or install a smoke alarm with a pause button.

    + If a smoke alarm sounds during normal cooking, press the pause button if the smoke alarm has one. Open the door or window or fan the area with a towel to get the air moving. Do not disable the smoke alarm or take out the batteries.

    + Treat every smoke alarm activation as a likely fire and react quickly and safely to the alarm.

Chapter 7

• •

Technology and Cooking Fires

Several technological solutions have been proposed to address the cooking fire problem. Some of these involve preventing ignition, while others deal with methods of extinguishment. Although considerable research has been conducted on these technologies, few described here have yet been approved by standardization or certification organizations, and most or all still need additional evaluation. However, many show promise and would address significant portions of the home cooking fire problem.

Arthur D. Little's report for CPSC evaluated possible prevention technologies.

Arthur D. Little, Inc.'s 2001 report to the CPSC on technologies to address surface cooking fires evaluated a variety of technologies in terms of the cooking process, cooking time, consumer impact, maintenance requirements, cookware required, cooktop performance after actuation of safety system, reliability, durability, safety, applicability, availability, installation, serviceability, percent of surface cooking incidents and new equipment addressed, degree of mitigation, difficulty of system verification, potential for false results, resetability, and compliance with standards.[27] This report found that:

• Most cooktop fires could be mitigated by requiring attention or preventing the ignition of cooking materials.

Approximately 65 to 70 percent of surface cooking fires could be mitigated by requiring

someone to pay attention to the cooking using technologies such as timers and motion sensors. An additional 72 percent of the gas cooktop and 77 percent of the electric cooktop fires could be mitigated by preventing cooking materials from igniting using technologies such as temperature sensors.

• Different equipment configurations pose challenges.

While gas units may have open burners with pilot lights or electronic ignition, or have sealed burners, electric units may have glass ceramic, smooth-top surfaces or open coil burners. In addition, roughly 10 percent of ranges do not have hoods, eliminating hood installation of safety equipment as an option for these kitchens. The varied types of cooking surfaces in use make different prevention technology options more or less feasible.

• Different combinations of temperature sensor, fusible link, timer, and motion and power sensor technologies could be used.

USFA explored hood suppression systems and kitchen-only sprinklers.

A USFA study examined the possibility of using inexpensive "active" fire protection systems that could be retrofitted easily and be effective in extinguishing a typical kitchen fire.[28] The goal was to find effective systems costing $200 or less. The study examined the following three types of systems: 1) a wet chemical system installed under the range exhaust hood, 2) a dry chemical system installed under the range exhaust hood, and 3) a single automatic fire sprinkler in the kitchen.

The kitchens of an abandoned apartment building were used to test these systems. Both hood systems had been effective in laboratory tests and were effective on cooking oil fires. The one low-flow residential automatic fire sprinkler in the kitchen also was effective in controlling a cooking oil fire as well as a countertop appliance fire, even when the cabinets shielded the fire from the sprinkler and the fire had spread to the walls and cabinets before the sprinkler operated. However, none of the systems shut off the electricity or gas to the stove.

Both range exhaust hood systems tested were found to be inexpensive. In addition, costs to install one single sprinkler would be manageable if the domestic water supply could be used. However, a homeowner would need to be sure that water pressure and flow are sufficient if the one single sprinkler is to be used effectively.

All that being said, a comprehensive cost-effectiveness assessment of kitchen-only sprinklers and hood suppression systems has yet to be completed and would need to compare the lower cost with the lower benefit, factoring in reliability (which has historically been worse for hood suppression systems than for water-based sprinklers in typical commercial installations) and the potential for fire spread outside the protected area by fires that begin away from the primary cooking area and countertops (such as some fires due to smoking, heating equipment, or electrical distribution equipment).

Summary Discussion

Technology can be used to prevent ignition or to mitigate the effects if a fire should occur. For example, technological systems that limit a stove's heat or shut off the cooking equipment before or when a fire occurs have some obvious advantages. While it is imperative that individuals adhere to safe cooking behaviors, technology may be the best long-term solution to dealing with the cooking fire problem. However, any technological solution must be proven effective and applicable to all types of cooking. In addition, to gain wide market acceptance, it must be inexpensive.

Implications for Behavioral Strategies

At this time, there are no implications for behavioral strategies on using technology to address the cooking fire problem. However, innovative products like those discussed in this chapter may provide greater cooking safety without requiring so much care and continuous alertness by cooks and residents.

Chapter 8

● ●

Other Cooking, Food, and Hot Beverage Burns

Cooking and hot food or beverages account for large numbers of burns that are not caused by fires, particularly thermal burns from contact with hot equipment and containers, and scald burns. Fire and life safety messages and behavioral strategies targeted on cooking fires should be complemented by messages and strategies that target these types of burns.

U.S. emergency rooms treated 19,600 thermal burns caused by ranges in 2004.

Many individuals are injured by hot cooking equipment, dishes, food, or liquids while cooking, while someone else is cooking, or while consuming cooked or heated food or beverages. Figure 33

Figure 33. Cooking-Equipment-Related Thermal Burns Seen in U.S. Hospital Emergency Rooms in 2004

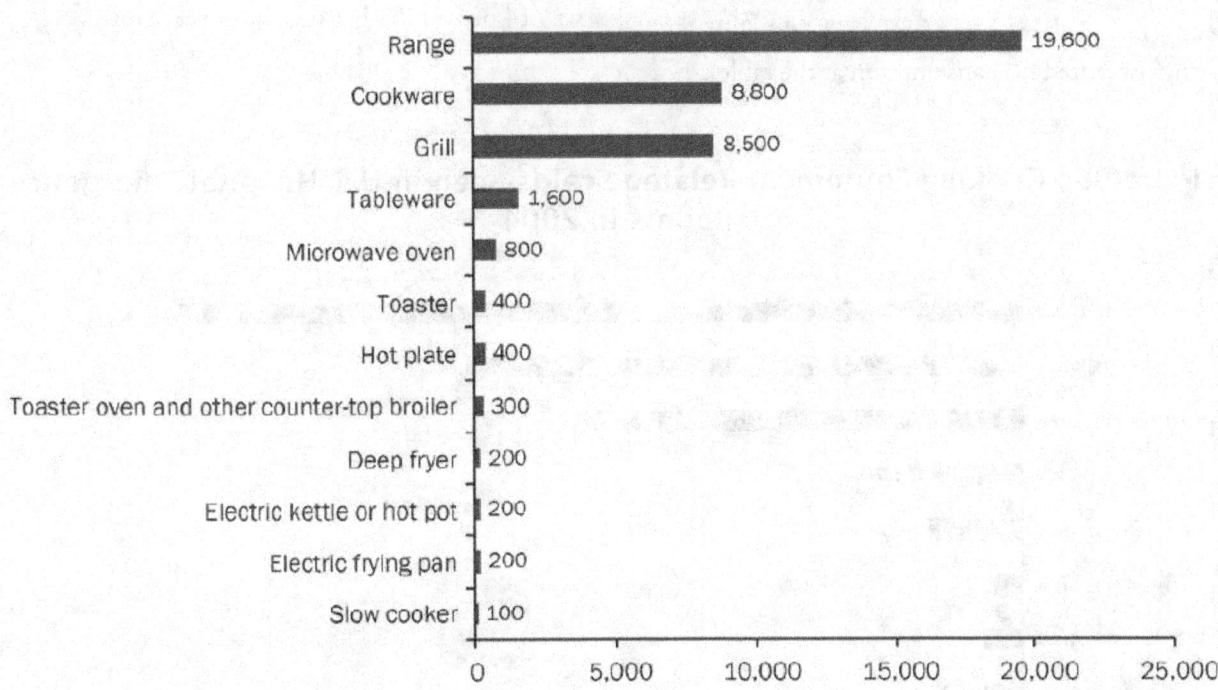

55

Figure 34. Range-Related Thermal Burns per Million Population Seen in U.S. Hospital Emergency Rooms in 2004, by Age Group

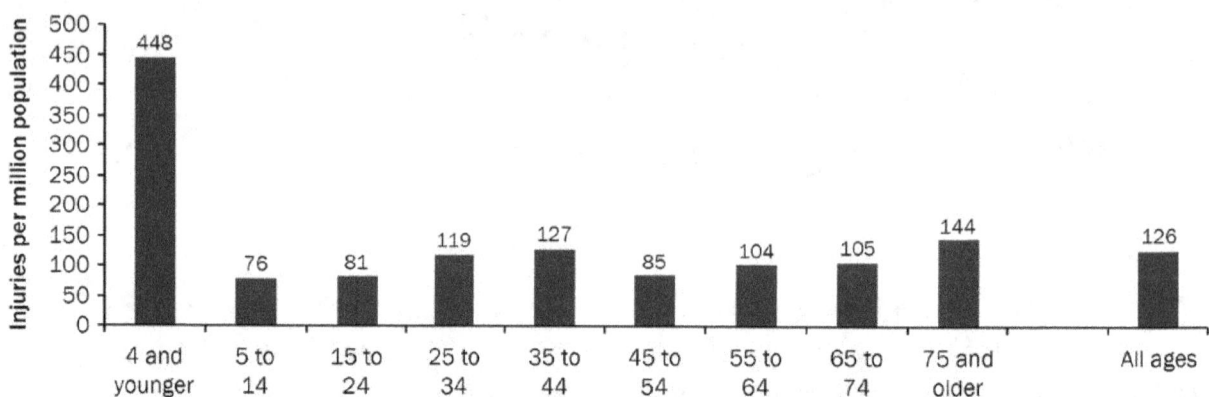

shows that, according to the CPSC's NEISS, thermal burns from ranges sent an estimated 19,600 people to U.S. emergency rooms in 2004. Thermal burns include both flame burns and contact burns. Figure 33 also shows that 8,800 thermal burns were caused by cookware (i.e., containers for food or liquid while it is being heated, such as a pot, pan, or unpowered coffee pot or tea pot), 8,500 were caused by grills, and 1,600 were caused by tableware (i.e., containers for food or drink while it is being presented for consumption at the table).

Preschoolers were at the highest risk of thermal cooking burns from ranges.

Figure 34 shows that children under 5 faced a risk of range thermal burns that was 3.6 times that of the general population. Adults 75 and older were the only other age group with an above-average risk of these burns as their rate of thermal burn injury was 14 percent higher than the overall rate.

Figure 35. Cooking-Equipment-Related Scalds Seen in U.S. Hospital Emergency Rooms in 2004

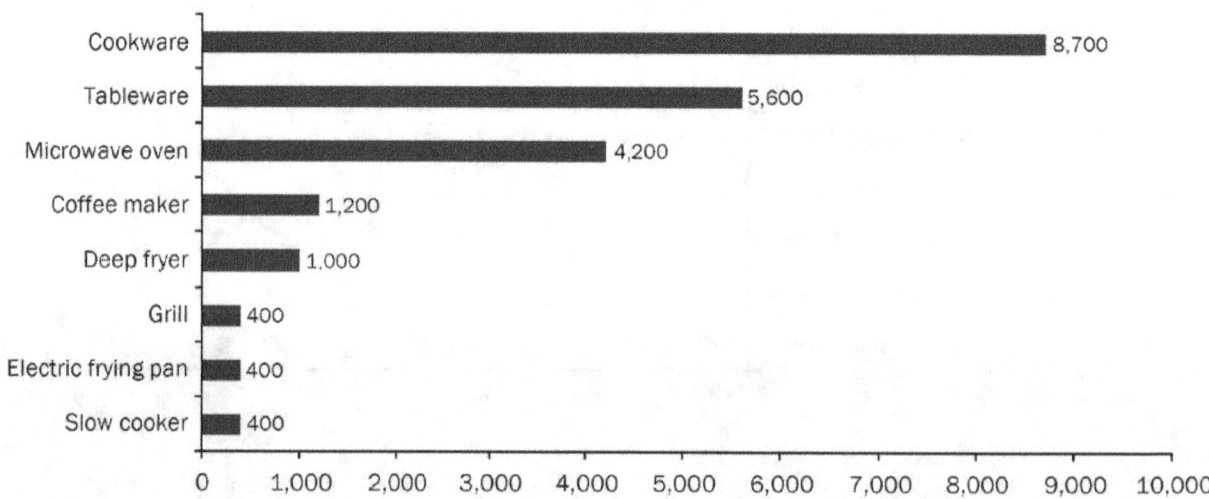

Figure 36. Cooking-Related Scald Injuries per Million Population Seen in U.S. Hospital Emergency Rooms in 2004, by Age Group

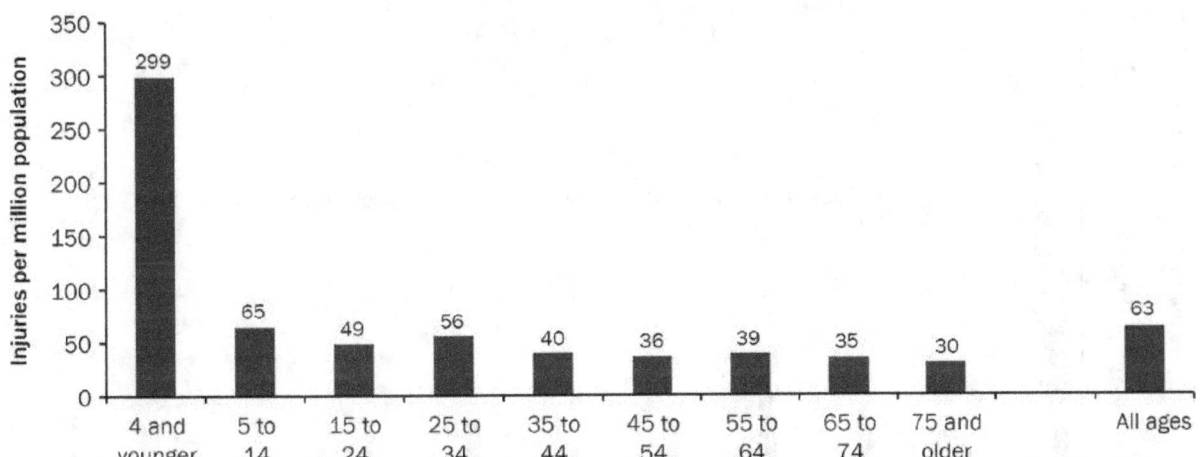

U.S. emergency rooms treated 8,700 cookware and 5,600 tableware scalds in 2004.

Scalds also are associated with cooking activities. A scald involves damage from steam or hot water, while a thermal burn involves damage due to contact with a flame or hot object. Figure 35 shows that cookware was involved in 8,700 scalds seen in hospital emergency rooms in 2004; tableware was involved in 5,600 scalds; microwave ovens were involved in 4,200 scalds; coffeemakers were involved in 1,200 scalds; and deep fryers were involved in 1,000 scalds. Some scalds occurred when a hot liquid was poured from cookware to tableware (i.e., teapot to teacup). Others occurred when a container was knocked over, bumped, dropped, or pulled over by the victim or someone else.

Preschoolers also were at increased risk from cooking-related scalds.

Figure 36 shows that children under 5 suffered almost five times the rate of cooking-related scald injuries as the population as a whole. Unlike thermal burns, older adults were not at increased risk from these burns.

Tableware posed almost twice the risk of preschooler scalds as cookware, and three times the risk as microwave ovens.

Figure 37 shows that the greatest threat of cooking-related scald burns to young children came from tableware such as cups or bowls. Children under 5 had the highest rate of scald injury from all three of the leading types of cooking-related equipment shown, including cookware, tableware, and microwave ovens.

Scalds accounted for two-thirds of the home cookware burns in children under 6.

A study of kitchen scalds and thermal burns in children under 6 years old reviewed data from CPSC's NEISS on emergency room visits from 1997 to 2002 for home burns incurred by children under 6 associated with nonelectric metal and nonmetal cookware as well as nonspecified cookware.[29] Sixty-six percent of the burns were scalds and 34 percent were thermal burns. The thermal burns most commonly resulted from touching hot pans.

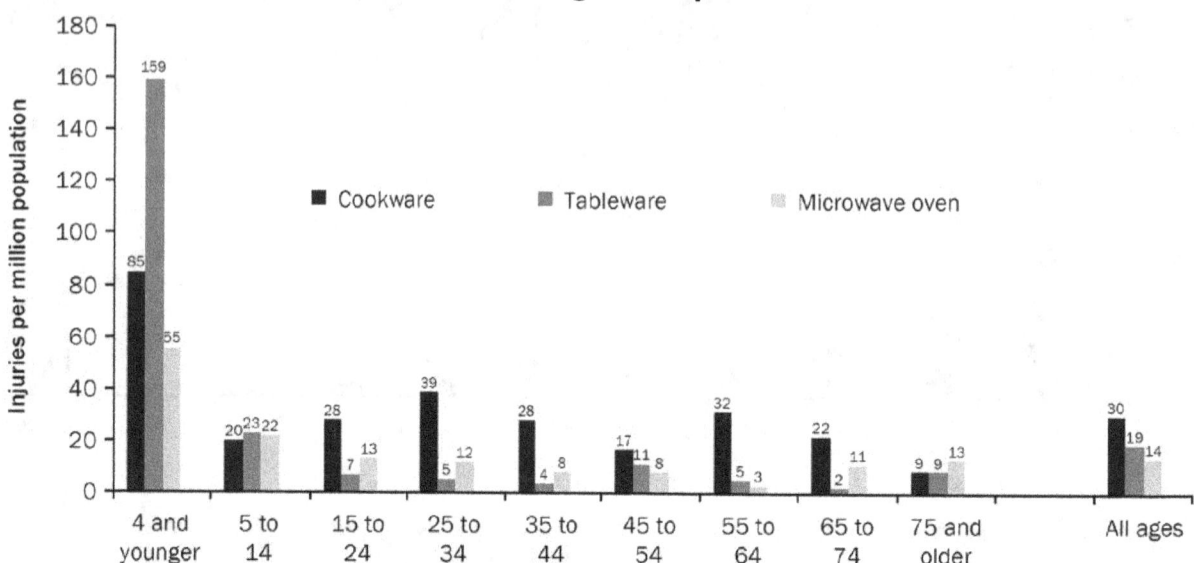

Figure 37. Cooking-Related Scald Injuries per Million Population Seen in U.S. Hospital Emergency Rooms in 2004, by Type of Product and Age Group

Children between one and two had the largest share of cookware scalds and thermal burns.

Figure 38 shows that the peak age for both types of burns in this group was 1 year. Two-thirds of the burns in this study were incurred by children under 3 years old.[29] Boys incurred 58 percent of the scalds and 55 percent of the thermal burns seen in this population.

Figure 39 shows that half of the scald injuries to children under 6 years of age resulted from children pulling a pot down or grabbing, knocking over, or spilling a pot. One-year-old children faced higher risks from these scenarios than did the other children. These children also faced a higher risk of scalds from pot contents splashing and from putting their hand in a pot. Five-year-olds faced a higher risk of colliding with a pot or person holding a pot.[29]

The author notes that the typical U.S. stove is 36 inches high and that average 1- and 2-year-old children can reach items on the front burners. A 2-year-old also could stretch more than halfway across a table to reach something of interest. The author also cited Shubert et al.'s 1990 finding that typical kitchen counters may be out of view of young children, but not out of reach.[29]

Stove burns to persons under 20 were more severe than burns from microwave ovens.

Using 1986 to 1990 data from CPSC's Injury Information Clearinghouse, Powell and Tanz compared burns associated with microwave and conventional stoves incurred by children (under 20 years old).[30] During the 5-year period, microwave burns were estimated at 5,160, while conventional stove burns were estimated at 41,198.

For microwave ovens, the mean age of child burn victim was 7.6 years, the median was 6 years old. One-quarter of the victims were under 3 years of age. Scalds accounted for 95 percent of the microwave burns. Exploding foods, such as eggs, accounted for 16 percent of the scalds. None of the

Figure 38. Cookware Scald and Thermal Burns to Children Five and Under Seen in Hospital Emergency Rooms by Age

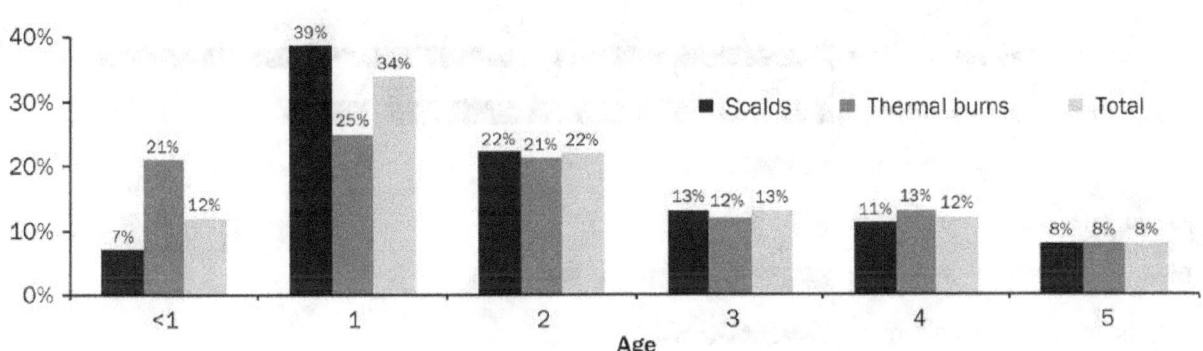

Figure 39. Cookware Scald Injury Patterns for Children Under Six Seen in Hospital Emergency Rooms

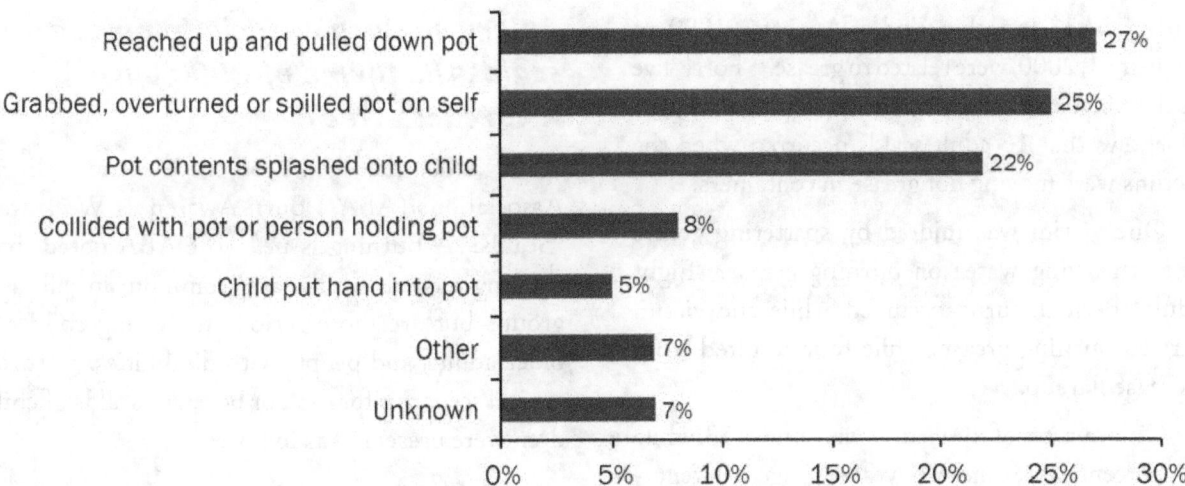

microwave burns exceeded 25 percent of total body surface area and none required hospitalization.

For conventional stoves, the mean age of child burn victims was 5.8 years; the median was 3 years old. Forty-five percent of the burns were incurred by children under 3 years of age. Thermal burns accounted for 74 percent of the stove burns. Seven percent of the stove burns exceeded 25 percent total body surface area and 5 percent of the child burn victims were hospitalized. The authors

conclude that child burns associated with microwaves are less frequent and less severe than those caused by stoves, and recommend that burn prevention emphasize stove hazards.

Grease burns accounted for 9 percent of acute burn admissions to Still Burn Center.

Sixty, or 9 percent, of the acute burn admissions to the Still Burn Center in Augusta, Georgia,

Figure 40. Causes of Adult Grease Burns at Joseph M. Still Burn Center: August 1, 1999-August 31, 2000

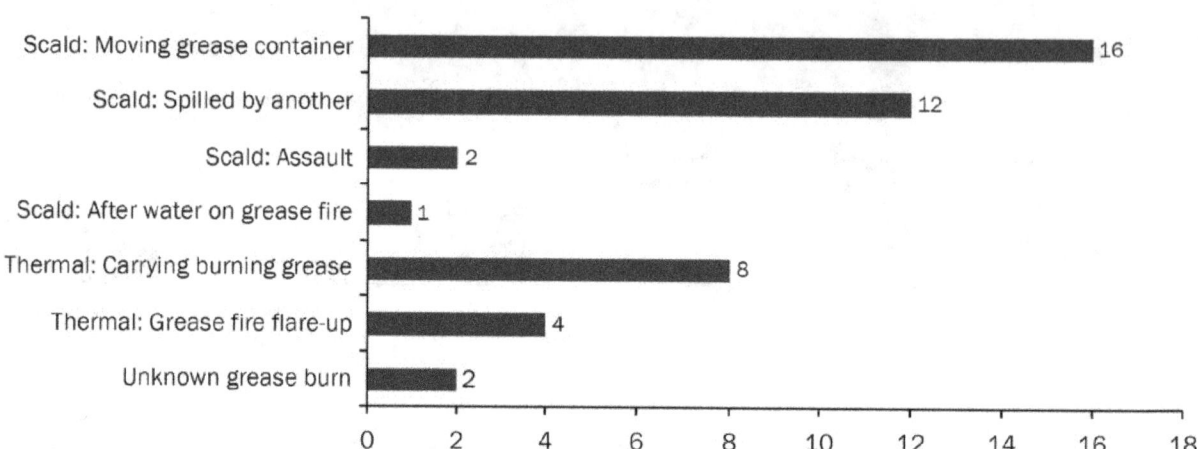

during the 13-month period of August 1, 1999, to August 31, 2000, were related to grease.[31] Forty-five of the 60 grease burn patients were adults. Figure 40 shows that 16 adult scalds occurred when the victims were moving hot grease in containers.

One victim was injured by spattering grease after throwing water on burning grease. Eight adult thermal burns occurred while the victim carried burning grease, while four resulted from a grease flareup.

One-quarter of the victims were under 18, while 10 percent were under 4 years of age. Twenty-percent of the grease burns in this study were caused by deep fryers. Seventy-eight percent of the adult grease burns and 14 of the 15 juvenile grease burns were scalds.

The authors note that improper supervision was often a factor in the child burns. Two toddlers pulled the cords on deep fryers. One youngster was scalded by pulling over a hot grease container. Another pulled a towel out from under a deep fryer, pulling the fryer at the same time. An additional one was scalded while carrying hot grease, while another was injured while carrying a pot of burning grease.

American Burn Association made scalds the theme of 2000 Burn Awareness Week.

In 2000, the theme of the American Burn Association's (ABA's) Burn Awareness Week was "Scalds: A Burning Issue." The ABA noted that cooking-related scalds are common in all age groups but are more serious for young children, older adults, and people with disabilities.[32] Three typical scenarios for food or beverage scalds of children were presented as follows:

1) A child pulls a pan or other container of hot liquid off the range or counter.

2) A toddler collides with an adult carrying these items.

3) A toddler pulls on a tablecloth and spills the food or drink that had been on the table.

Coffee and other hot beverages are normally served at 160 to 180 °F (71 to 82 °C). Water boils at 212 °F (100 °C). Frying is generally done at 300 °F (148 °C), while deep frying temperatures can reach 500 °F (260 °C). Citing a 1947 reference from Mortiz and Herriques, they note that third-degree burns can result from exposure to water at:

+ 155 °F (68 °C) for one second.

+ 140 °F (60 °C) for five seconds.

+ 133 °F (56 °C) for 15 seconds.

These points can help illustrate to lay audiences how quickly a very serious scald burn can occur. The temperature points chosen are lower than the normal serving temperatures for hot coffee. The ABA also provided detailed safety tips for preventing food and beverage related-scalds as well as microwave oven scalds. Tips from ABA can be found in Appendix D.

Drago cautions that many recommendations have been made for years with little effect.

Dorothy Drago notes that many childhood scald prevention strategies have been ineffective at reducing these injuries. Active intervention, such as turning pot handles inward, putting hot drinks in the middle of tables so toddlers cannot reach them, and removing tablecloths, were recommended from 1977 on but have had little effect. She cited Van Rijn et, al.'s 1991 study on behavioral risk factors for burn injuries, writing, "For most parents in the van Rijn study, the reason that they did not implement a desired safety behavior was that they were not familiar with it, the behavior was not habitual, and they were not able to resist the pressure of others. Parents who did implement a safety behavior were able to associate the behavior with the prevention of burn injuries more than those who did not make the connection." If parents were more aware of child development and skill acquisition, they might be better able to see the risk of kitchen burns and to prevent them.[29]

Summary Discussion

Many foods and beverages are served customarily at temperatures that can cause a third-degree burn in just 1 second. Often the fire safety community has overlooked non-fire burn injuries associated with cooking, although they are related to cooking fires and share some common behavioral mitigation messages. For example, cooking oil and grease fires have long been a concern to the fire service as they make up a significant portion of cooking fires. Cooking oil is also a frequent source of scald burns.

The risk of thermal burns or scalds from cooking equipment, cookware, tableware or hot foods or beverages is noticeably high, especially for young children. In order to increase the likelihood of the implementation of certain behaviors, the fire service community has a shared responsibility to educate people, especially parents of young children, on how and why specific behaviors would prevent devastating burns and scalds and stress the importance of being aware of childhood development and skill acquisition.

Behavioral Strategies

The following specific messages arising from this chapter address protecting children from scalds and burns, preventing and treating scalds and burns, using microwave ovens safely, and preventing cooking fires in general:

1. Protect children from scalds and burns.

   + Young children are at high risk of being burned by hot food and liquids. Keep children away from cooking areas by enforcing a "kid-free zone" of 3 feet (1 meter) around the stove.

   + Keep young children at least 3 feet (1 meter) away from any place where hot food or drink is being prepared or carried. Keep hot foods and liquids away from table and counter edges.

   + When young children are present, use the stove's back burners whenever possible.

   + Never hold a child while cooking, drinking, or carrying hot foods or liquids.

+ Teach children that hot things burn.

+ When children are old enough, teach them to cook safely. Supervise them closely.

2. Prevent and treat scalds and burns.

+ To prevent spills due to overturn of appliances containing hot food or liquids, use the back burner when possible and/or turn pot handles away from the stove's edge. All appliance cords need to be kept coiled and away from counter edges.

+ Use oven mitts or potholders when moving hot food from ovens, microwave ovens, or stovetops. Never use wet oven mitts or potholders as they can cause scald burns.

+ Replace old or worn oven mitts.

+ Treat a burn right away, putting it in cool water. Cool the burn for 3 to 5 minutes. If the burn is bigger than your fist or if you have any questions about how to treat it, seek medical attention right away.

3. Install and use microwave ovens safely.

+ Place or install the microwave oven at a safe height, within easy reach of all users. The face of the person using the microwave oven should always be higher than the front of the microwave oven door. This is to prevent hot food or liquid from spilling onto a user's face or body from above and to prevent the microwave oven itself from falling onto a user.

+ Never use aluminum foil or metal objects in a microwave oven. They can cause a fire and damage the oven.

+ Heat food only in containers or dishes that are safe for microwave use.

+ Open heated food containers slowly away from the face to avoid steam burns. Hot steam escaping from the container or food can cause burns.

+ Foods heat unevenly in microwave ovens. Stir and test before eating.

References

1. Schreoder, Tom, and Kimberly Ault. *The NEISS Sample Design And Implementation: 1997 to Present.* U.S. Consumer Product Safety Commission, June 2001. Online at http://www.cpsc.gov/neiss/2001d011-6b6.pdf

2. Hall, John. *Home Cooking Fire Patterns and Trends.* Quincy: National Fire Protection Association, July 2006.

3. U.S. Fire Administration, National Fire Data Center. *National Fire Incident Reporting System 5.0 Complete Reference Guide.* Jan. 2006. Online at http://nfirs.fema.gov/documentation/reference/

4. Audits and Surveys—Government Research Division. *1984 National Sample Survey of Unreported Residential Fires, Final Technical Report*, prepared under contract no. C-83-1239 for U.S. Consumer Product Safety Commission, June 1985.

5. Miller, David. *Estimates of Fire Injuries Treated in Hospital Emergency Departments: July 2002-June 2003.* Washington: U.S. Consumer Product Safety Commission, Division of Hazard Analysis, Directorate of Epidemiology, 2005. Accessed on-line at http://www.cpsc.gov/library/neissfire.pdf on Feb. 28, 2006.

6. Hamrick, Karen, and Kristina J. Shelley. "How Much Time Do Americans Spend Preparing and Eating Food?" *Amber Waves*, Nov.

2005. Online at http://www.ers.usda.gov/AmberWaves/November05/DataFeature/

7. National Association of State Fire Marshals (NASFM) Cooking Fires Task Force and Association of Home Appliance Manufacturers (AHAM) Safe Cooking Campaign. *Ten-Community Study of the Behaviors and Profiles of People Involved in Residential Cooking Fires: Executive Summary.* July 1996. Online at http://66.220.163.24/cooksafe/10cityrpt.cfm

8. Smith, Linda, Ron Monticone, and Brenda Gillum. *Range Fires, Characteristics Reported in National Fire Data and a CPSC Special Study,* Washington: U.S. Consumer Product Safety Commission, Division of Hazard Analysis, Directorate of Epidemiology, 1999. Online at http://www.cpsc.gov/LIBRARY/FOIA/Foia99/os/range.pdf

9. Motz, George, Linzee Liabraaten, and Heather White. "The Design of Everyday Stoves: User Centered Design in the Kitchen." Class paper for EDIT 797: Performance Centered Design at George Mason University, Nov. 25. 2003. Online at http://immersion.gmu.edu/portfolios/gmotz/PerformanceCenteredStoves.pdf

10. U.S. Census Bureau. *Statistical Abstract of the United States: 2006,* 125th ed. Washington: Author, 2005. Table 961, "Heating Equipment and Fuels for Occupied Units: 1995 to 2003."

11. Lemoff, Ted. Gas Range Questions, personal communication, Jan. 20, 2006.

12. National Fire Protection Association. *Cooking Safety: Turkey Fryers.* Accessed online at http://www.nfpa.org/itemDetail.asp?categoryID=282&itemID=27798&URL=Research%20&%20Reports/Fact%20sheets/Home%20safety/Cooking%20safety on July 20, 2006.

13. Underwriters Laboratories Inc. *Product Safety Tips: Turkey Fryers.* Online at http://www.ul.com/consumers/turkeys.html on Sept. 5, 2006.

14. Beaufort, South Carolina, Fire Department. *Beaufort Firefighters Urge Caution when Frying Turkeys.* Press release received in e-mail correspondence from Daniel Byrne on Nov. 29, 2005.

15. Scottsdale, Arizona, Fire Department. *Deep Fry Your Turkey Safely.* Accessed online at http://www.scottsdaleaz.gov/safety/Fire/turkeyfryersafety.asp on July 20, 2006.

16. Rowe, William. *Butane Fueled Table Top Cooking Appliances: Staff Project Report,* U.S. Consumer Product Safety Commission, May 2003. Online at http://www.cpsc.gov/library/foia/foia04/os/cooking.pdf

17. Consumer Product Safety Commission. *Aluminum Cookware Can Melt and Cause Severe Burns.* CPSC Document #5015. Accessed at http://www.cpsc.gov/cpscpub/pubs/5015.html on May 18, 2006.

18. National Fire Protection Association. *Grilling.* Accessed online at http://www.nfpa.org/itemDetail.asp?categoryID=298&itemID=18346&URL=Research%20&%20Reports/Fact%20sheets/Seasonal%20safety/Grilling on July 20, 2006.

19. American Burn Association. *2002 Burn Awareness Week Campaign Kit: Summer Recreational and Camping Burn Prevention.* Online at http://www.ameriburn.org/Preven/2002Prevention/BurnPreventionKit2002(Final).pdf

20. Key Research and Marketing, Ltd. *New Zealand Fire Service Bay-Waikato Fire Region Kitchen Fire Research, Summary of Findings.* Oct. 1998. Online at http://baywaikato.fire.org.nz/research/pdf/kitchen.pdf

21. Rohr, Kimberly D. *Products First Ignited in Home Fires*. Quincy: National Fire Protection Association, Apr. 2005.

22. National Energy Assistance Directors' Association. *2005 National Energy Assistance Survey: Final Report*. Sept. 2005. Online at http://www.neada.org/comm/surveys/NEADA_2005_National_Energy_Assistance_Survey.pdf

23. *American Heritage Dictionary of the English Language*, 3rd ed. New York: Houghton Mifflin Company, 1996.

24. Ahrens, Marty. *U.S. Fires in Selected Occupancies*. Quincy: National Fire Protection Association, Mar. 2006.

25. Harris Interactive Market Research. *Fire Prevention Week Survey* conducted for National Fire Protection Association, 2004. Online at http://www.nfpa.org/assets/images/Public%20Education/FPWsurvey.pdf

26. Smith, Charles L. *Smoke Detector Operability Survey—Report on Findings*, Bethesda: U.S. Consumer Product Safety Commission, Nov. 1993. Online at http://www.cpsc.gov/library/foia/foia01/os/operable.pt1.pdf

27. Arthur D. Little, Inc. *Technical, Practical and Manufacturing Feasibility of Technologies to Address Surface Cooking Fires, Final Report to United States Consumer Product Safety Commission*, May 2001. Online at http://www.cpsc.gov/library/foia/foia01/brief/ranges.pt1.pdf.

28. U.S. Fire Administration. *Localized Fire Suppression Systems*. Accessed online at http://www.usfa.dhs.gov/research/dsn/nist11.shtm on July 5, 2006.

29. Drago, Dorothy A. "Kitchen Scalds and Thermal Burns in Children Five Years and Younger." *Pediatrics*, 115, No. 1 (2005): 10-16. Accessed online at http://pediatrics.aappublications.org/cgi/reprint/115/1/10 on March 15, 2006.

30. Powell. E.C., and R.R.Tanz. , "Comparison of Childhood Burns Associated with Use of Microwave Ovens and Conventional Stoves." *Pediatrics*, 1993, 91, no. 2, (1993): 344-349. Abstract only, accessed at http://pediatrics.aappublications.org/cgi/content/abstract/91/2/344 on May 19, 2006.

31. Fiebiger, Barbara, Faye Whitmire, Edward Law, and Joseph Still. "Causes and Treatments of Burns from Grease." *Journal of Burn Care and Rehabilitation*, (2004), 25:374-376.

32. American Burn Association. Scalds: *A Burning Issue—A Campaign Kit for Burn Awareness Week 2000*. Online at http://www.ameriburn.org/Preven/2000Prevention/Scald2000PrevetionKit.pdf

Appendix A

How National Estimates Statistics Are Calculated

The statistics in this analysis are estimates derived from the U.S. Fire Administration's (USFA's) National Fire Incident Reporting System (NFIRS) and the National Fire Protection Association's (NFPA's) annual survey of U.S. fire departments. NFIRS is a voluntary system by which participating fire departments report detailed factors about the fires to which they respond. Roughly two-thirds of U.S. fire departments participate, although not all of these departments provide data every year.

The strength of NFIRS is that it provides the most detailed incident information of any national database not limited to large fires. NFIRS is the only database capable of addressing national patterns for fires of all sizes by specific property use and specific fire cause. NFIRS also captures information on the extent of flame spread and automatic detection and suppression equipment. For more information about NFIRS visit http://www.nfirs.fema.gov/

NFPA conducts an annual stratified random sample survey of fire departments, which captures a summary of fire department experience on a larger scale. The NFPA survey is based on a stratified random sample of roughly 3,000 U.S. fire departments (or just over one of every 10 fire departments in the country). The survey includes the following information: (1) the total number of fire incidents, civilian deaths, and civilian injuries, and the total estimated property damage (in dollars) for each of the major property-use classes defined by the NFPA 901 Standard; (2) the number of onduty firefighter injuries, by type of duty and nature of illness; and (3) information on the type of community protected (e.g., county versus township versus city) and the size of the population protected, which is used in the statistical formula for projecting national totals from sample results.

The NFPA survey begins with the NFPA Fire Service Inventory, a computerized file of about 30,000 U.S. fire departments. The survey is stratified by size of population protected to reduce the uncertainty of the final estimate. Small rural communities protect fewer people per department and are less likely to respond to the survey, so a large number must be surveyed to obtain an adequate sample of those departments. (NFPA also makes follow-up calls to a sample of the smaller fire departments that do not respond, to confirm that those that did respond are truly representative of fire departments their size.) On the other hand, large city departments are so few in number and protect such a large proportion of the total U.S. population that it makes sense to survey all of them. Most respond, resulting in excellent precision for their part of the final estimate. The results of the survey are published in the annual report Fire Loss in the United States. To download a free copy of the report visit http://www.nfpa.org/assets/files/PDF/OS.fireloss.pdf

Projecting NFIRS to National Estimates

As noted, NFIRS is a voluntary system. Different States and jurisdictions have different reporting requirements and practices. Participation rates in NFIRS are not necessarily uniform across regions and community sizes, both factors correlated with frequency and severity of fires. This means NFIRS may be susceptible to systematic biases. No one at present can quantify the size of these deviations from the ideal, representative sample, so no one can say with confidence that they are or are not serious problems. But there is enough reason for concern so that a second database—the NFPA survey—is needed to project NFIRS to national estimates and to project different parts of NFIRS separately. This multiple calibration approach makes use of the annual NFPA survey where its statistical design advantages are strongest.

Scaling ratios are obtained by comparing NFPA's projected totals of residential structure fires, non-residential structure fires, vehicle fires, and outside and other fires, and associated civilian deaths, civilian injuries, and direct property damage with comparable totals in NFIRS. Estimates of specific fire problems and circumstances are obtained by multiplying the NFIRS data by the scaling ratios.

Analysts at the NFPA, the USFA and the Consumer Product Safety Commission (CPSC) have developed the specific analytical rules used for this procedure. "The National Estimates Approach to U.S. Fire Statistics," by John R. Hall, Jr., and Beatrice Harwood, provides a more detailed explanation of national estimates. A copy of the article is available online at http://www.nfpa.org/osds or through NFPA's One-Stop Data Shop.

Version 5.0 of NFIRS, first introduced in 1999, used a different coding structure for many data elements, added some property use codes, and dropped others. It also introduced incident type codes for certain confined structure fires, including confined cooking fires, confined chimney fires, confined fuel burner fires, confined incinerator and compactor fires, and contained or confined trash fires. Very limited causal information is required for these incidents.

Note that percentages are calculated from unrounded values, and so it is quite possible to have a percentage entry of up to 100 percent, even if the rounded number entry is zero.

Fires Originally Collected in NFIRS 5.0 by Year

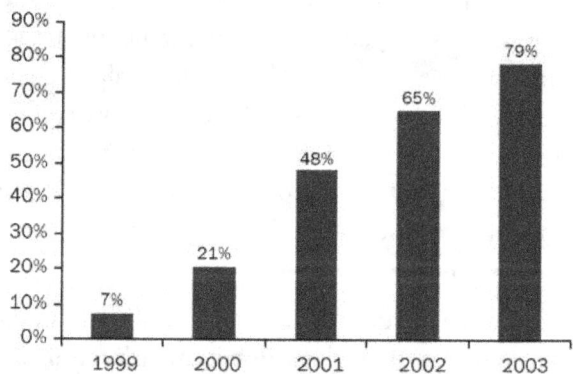

Appendix B

Existing Educational Messages Related to Civilian Firefighting for Cooking Fires

Advice on civilian firefighting from a variety of sources, including national organizations or agencies such as the NFPA, USFA, CPSC, and AHAM; local fire departments, groups; or institutions that have an interest in general safety; burn prevention specialists; and popular media are compiled here. While this material was identified during the project's literature review, there is no intent to suggest that these are the only sources with advice published or posted on civilian firefighting for cooking fires. There is also no intent to endorse or recommend all these statements of advice. In fact, there is considerable contradiction among the statements, which is the main point of this section.

Given that many people try to fight cooking fires themselves, consistent, sound, and realistic advice on how to fight these fires is needed.

The authors of the NASFM and AHAM 10-community study noted that, although public fire educators have not yet reached consensus on exactly what to teach people who have cooking fires (get out versus stay and fight), the question must be addressed.[1] Leaving the area and immediately calling the fire department when a cooking fire occurs may be the safest course of action. It is possible, however, that if people had not dealt with the unreported cooking fires on their own, these fires could have resulted in far more losses. In addition, the tendency of many people to try to fight a fire in their home suggests that they should also be given information on how to decide whether to attempt to stay as well as proper strategies for extinguishing or containing small cooking fires.

Jim Crawford, a Vancouver, Washington, fire marshal writing in Fire Rescue Magazine, said that self-firefighting messages have warned against using water on grease fires as well as against trying to use sugar, flour, or baking powder as extinguishing agents. Instead, fire extinguishers have been encouraged for cooking fires, although sometimes the improper use of an extinguisher presence has caused a fire to spread. He also said that the consensus seemed to advocate covering the pan and lowering the heat as the most effective method, and simplest and best choice, despite a possible risk of being burned.[2]

AHAM and its associates advise a lid for grease fires and baking soda for other food fires.

In the event of a kitchen cooking fire, the AHAM's Recipe for Safer Cooking suggests to "Call the fire department immediately... Slide a pan lid over flames to smother a grease or oil fire, then turn off the heat and leave the lid in place until the pan cools. Never carry the pan outside. Extinguish other food fires with baking soda. Never use water or flour on cooking fires. Keep the oven door shut and turn off the heat to smother an oven or broiler fire. Keep a fire extinguisher in the kitchen. Make sure you have the right type of training."[3] The USFA also advises smothering a pan fire with a lid and cautions against using water on a grease fire.[4]

NFPA cautions that potholders or mitts should be used when putting a lid on a fire.

NFPA promotes the following messages on its cooking fact sheet:

+ "Always keep a potholder, oven mitt and lid handy. If a small fire starts in a pan on the stove, put on an oven mitt and smother the flames by carefully sliding the lid over the pan. Turn off the burner. Don't remove the lid until it is completely cool. Never pour water on a grease fire and never discharge a fire extinguisher onto a pan fire, as it can spray or shoot burning grease around the kitchen, actually spreading the fire.

+ If there is an oven fire, turn off the heat and keep the door closed to prevent flames from burning you and your clothing.

+ If there is a microwave fire, keep the door closed and unplug the microwave."[5]

For grease or cooking oil fires specifically, NFPA advises:

+ "If the oil catches fire, wearing an oven mitt, immediately, but carefully, slide a lid over the pan to smother the fire. Turn off the burner and slide the pan off the heat source. Keep the pan covered until the oil cools to prevent it from starting again.

+ If the oil has overflowed from the pan and ignites, get everyone out of the home and call the fire department from outside.

+ Never use water to extinguish a cooking oil fire."[6]

The Casper, Wyoming, Fire Department discourages the use of salt or baking soda on grease fires.

The Casper, Wyoming, Fire Department developed online material on grease fire safety after becoming concerned that they were repeatedly hearing "the misconception that you can effectively extinguish a grease fire using salt, baking soda, or water."[7] They note that salt and baking soda will extinguish a fire if applied long enough to cover and smother the fire or fuel. However, reaching over the fire to get these materials is highly dangerous. Because of the high flames and quantity of smoke, pouring an adequate amount of salt or baking soda would be dangerous. Standing away from the fire and throwing salt or baking soda on it is not considered practical because of the amount and time that would be required to extinguish the fire. Flour could make a fire worse. Although baking soda and salt are cheaper than fire extinguishers, fire extinguishers are inexpensive, and should be mounted near an exit so that escape is possible. Carrying a burning pan to the sink and using water is wrong for two reasons. Dropping a burning pan can spread the fire or spill burning grease. Adding water will make the burning grease explode. A wet towel was not advised because of the time involved in getting one, and the fact that water and grease don't mix.

In the event of a grease fire, the Casper Fire Department recommends using an ABC fire extinguisher, the pan's lid, or a noncombustible item such as a cookie sheet. A lid or cookie sheet should be held like a shield when approaching the fire and gently placed. The heat should then be turned off. The most important point is to know when a fire is too big for an occupant to fight. A grease fire that has spread to the cabinets or the structure is too big. The fire department should be called even if the occupant has successfully extinguished the fire to ensure that the fire is out and not smoldering in the walls. The fire department's report also is useful in insurance claims. People also are cautioned to stay out of the smoke before and after the fire is out.

The Maine Farm Safety Fact Program mentions baking soda as an option for pan fires.

The University of Maine Cooperative Extension produces an extensive collection of fact sheets about various aspects of farm and home safety. One of their fact sheets addresses kitchen safety.[8] "When a fire occurs, assess the situation. Always give yourself a place to escape. If it is possible to safely turn off the electricity or gas feeding the fire, do so. If a pan is on fire, shut off the heat and tightly cover the fire with a lid. This should be done only if the fire is small. Never pour water on a pan fire involving grease, or try to carry it to the sink or outdoors.

If the above methods have failed, use a fire blanket, fire extinguisher or baking soda to put out the fire. When using a fire blanket, cover your hands with it and gently throw the blanket over the fire. Fire extinguishers should be sprayed at least one yard from the fire and aimed directly above the fire in the vapor area. Test the extinguisher before approaching the fire. Sweep it from side to side until the fire is out. Baking soda should be sprinkled or thrown onto the fire."

The Gainesville, Georgia, Fire Department mentions salt as an option for grease fires.

In their online fire prevention tips, the Gainesville, Georgia, Fire Department reminds readers that they should never try to put out a grease fire with water. Instead, they should "Smother the fire with a lid, use salt or other materials (fire extinguisher!) to extinguish the fire."[9]

Reader's Digest recommends salt for grease fires.

In its 31 Extraordinary Uses for Salt—in the Kitchen, the Reader's Digest Moderator advises storing a box of salt next to the range. If a grease fire occurs, the salt can be tossed on it to put out the fire. They also caution against using water. Salt also was considered useful when meat drippings lead to excessively high barbecue flames. Salt sprinkled on the coals can subdue the flames without creating a lot of smoke or cooling the coals. Adding salt to a pan before frying was said to prevent grease splatters and associated burns and mess.[10]

Other recommendations related to firefighting for grease fires:

- Vladimir Prpich of Monash University Residential Services in Australia advises that a fire extinguisher or fire blanket may be used to contain a cooking oil fire.[11]

- Hankins, Tang and Phipps recommend that labels on fryers and cooking oil bottles include the following warnings (among others):

  - "Don't transport oil that is hot or is ignited. Extinguish hot oil fires by placing a lid over the fire."

  - "Always have either a fire extinguisher or fire blanket on hand in the kitchen."

  - "Avoid the consumption of alcoholic beverages when using hot oil cooking appliances or deep-frying."[12]

- Gray, Cheng, and Pegg recommend school-based programs on what to do (and not to do) should cooking oil ignite. Students also should be taught to use "only properly designed cooking containers." The authors also recommend warning labels on hot oil cookware, cooking oil bottles, and foods used in deep fryers. They further recommend that homes have fire blankets or fire extinguishers.[13]

References

1. National Association of State Fire Marshals (NASFM) Cooking Fires Task Force and Association of Home Appliance Manufacturers (AHAM) Safe Cooking Campaign. *Ten-Community Study of the Behaviors and Profiles of People Involved in Residential Cooking Fires: Executive Summary.* July 1996. Online at http://66.220.163.24/cooksafe/10cityrpt.cfm

2. Crawford, Jim. "Beyond Baking Soda: New Technology May Make Kitchen Fires a Thing of the Past." *Fire Rescue Magazine,* Vol. 23, no. 12 Dec. 2005, pp. 78-79.

3. Association of Home Appliance Manufacturers, National Association of State Fire Marshals and National Safety Council. *Recipe for Safer Cooking.* Online at http://www.aham.org/ht/a/GetDocumentAction/i/588

4. United States Fire Administration. *Cooking Fires.* Accessed online at http://www.usfa.dhs.gov/media/quick_response/ffwf-11.shtm on Mar. 26, 2007.

5. National Fire Protection Association. *Tips for Safer Cooking.* Accessed online at http://www.nfpa.org/categoryList.asp?categoryID=387&URL=Learning/Public%20Education/Fire%20Prevention%20Week%202006/For%20the%20fire%20service/Tips%20for%20safer%20cooking on July 20, 2006.

6. National Fire Protection Association. *Cooking Safety: Cooking Oil Safety.* Accessed on line at http://www.nfpa.org/itemDetail.asp?categoryID=282&itemID=27800&URL=Research%20&%20Reports/Fact%20sheets/Safety%20in%20the%20home/Cooking%20safety on Jan. 24, 2007.

7. Casper, Wyoming, Fire Department. *Grease Fire Safety.* Accessed online at http://www.casperfire.com/fire_prevention/fp_grease-firesafety/grease_fire_safety.htm on June 5, 2006.

8. Cyr, Dawn La, and Steven B Johnson. *Kitchen Safety.* Maine Farm Safety Fact Program, Bulletin # 2314, 7/95. Accessed online on Apr. 25, 2006 from http://www.cdc.gov/nasd/docs/d000801-d000900/d000825/d000825.pdf

9. Gainesville, Georgia, Fire Department. *Fire Prevention Tips.* Accessed online at http://www.gainesville.org/citydepartments.firedepartment.firepreventiontips.asp on June 5, 2006.

10. Reader's Digest Moderator. *31 Extraordinary Uses for Salt—in the Kitchen.* Accessed online at http://www.gather.com/viewArticle.jsp?articleId=281474976757060 on June 5, 2006.

11. Prpich, Valdimir. *Hazard Alert—Cooking with Oil.* Monash University. Accessed online at http://www.mrs.monash.edu.au/on-campus/ohs-cooking-oil.html on June 6, 2006.

12. Hankins, Christopher L., Xia Qing Tang, and Alan Phipps. "Hot Oil Burns—A Study of Predisposing Factors, Clinical Course and Prevention Strategies." *Burns* 32 (2006) 92-96.

13. Gray, Katherine, Eddie Cheng, and Stuart Pegg.. "Hot Cooking Oil Burns: A 20-Year Experience." *Journal of Burn Care and Rehabilitation,* 2004; 25:205-210

Appendix C

Grilling Safety Messages from the American Burn Association's 2002 Burn Awareness Week

The following messages were issued by the American Burn Association (ABA) as part of the 2002 Burn Awareness Week campaign "Recreational and Camping Burn Prevention:"[1]

ABA propane grill messages:

+ Open the valve only a quarter to one-half turn before lighting.

+ Always shut off the valve to a fuel source when it is not in use.

+ Never start a gas grill with the lid of the grill closed. The propane or natural gas may accumulate inside and, when ignited, could explode and blow the lid off, causing injury.

+ Periodically, clean the Venturi tubes that displace the gas under the grill. When insects or debris block tubes, gas is forced out somewhere else within the system. Use the manufacturer's instructions for cleaning.

+ Have a BC-type fire extinguisher located in the grilling area.

ABA charcoal grill messages

+ After applying charcoal lighter fluid to the coals, wait a minute before lighting the coal. This allows the heavy concentration of explosive vapors to disperse.

+ Be careful not to spill any fluid on your clothing or in the area surrounding the grill.

+ Wear an insulated fire-retardant barbecue mitt when lighting coals.

+ If using a lighter to start the barbecue, remember the following:

  - Keep all lighters out of sight and out of reach of children.

  - Barbecue lighters (also called utility lighters or multipurpose lighters) are easy to use around the home and are convenient for camping. Among other things, they are often used to start barbecues and to light campfires, fireplaces, wood stoves, and candles.

  - Children find it easy to use these lighters. **Barbecue lighters are made to be used by adults and are NOT safe for children.** Even a small child can figure out how to pull the trigger. **Barbecue lighters are not toys!**

  - Do not leave a lighter outside. The weather can damage the plastic and the fuel inside may leak out or the lighter may break open.

  - BEFORE you use it, read all the instructions that come with the barbecue lighter.

  - Purchase barbecue lighters that say "child-resistant" on the package.

Reference

1. American Burn Association. 2002 Burn Awareness Week Campaign Kit: Summer Recreational and Camping Burn Prevention. Online at http://www.ameriburn.org/Preven/2002Prevention/BurnPreventionKit2002(Final).pdf

Appendix D

Scald Prevention Tips from the ABA's Scalds: A Burning Issue, A Campaign Kit for Burn Awareness Week 2000

The American Burn Association provided tips on scald prevention.

In their 2000 Burn Awareness week, the ABA provided detailed safety tips for preventing food- and beverage related-scalds as well as microwave oven scalds. The tips below come from the ABA's Scalds: A Burning Issue, A Campaign Kit for Burn Awareness Week 2000, found on their Web site.[1] These tips address scald prevention in the cooking and dining areas. Many of these, particularly those relating to protecting young children from hot cooking liquids, are also relevant to fire safety. Tips are also provided on preventing microwave oven and hot beverage scalds. Recommendations are also made for people with mobility impairments.

Because attention to scalds is still new to many in the fire safety community, the ABA tips are quoted here at length, for consideration in future changes to educational messaging for fire and life safety educators.

General scald prevention tips from the ABA.

+ Establish a safe area, out of the traffic path between the stove and sink, where children can safely play but still be supervised.

+ Place young children in high chairs or play yards a safe distance from counter or stovetops, hot liquids, hot surfaces, or other cooking hazards while preparing or serving food.

+ Child walkers are extremely dangerous and should never be allowed in kitchens or bathrooms. Infants in child walkers have increased mobility and height and can more easily come in contact with dangling cords and pot handles.

+ Provide safe toys for children, not pots, pans, and cooking utensils, to occupy a child's attention. Young children are unable to distinguish between a "safe" or "play" pan that they perceive as a toy and may reach for a pan on the stove.

+ Cook on back burners when young children are present.

+ Keep all pot handles turned back, away from the stove edge. All appliance cords need to be kept coiled and away from counter edges. Curious children may reach up and grab handles or cords. Cords also may become caught in cabinet doors causing hot food and liquids to spill onto you or others. The grease in deep fat fryers and cookers can reach temperatures higher than 400 degrees and cause serious burns in less than 1 second.

+ When removing lids from hot foods, remember that steam may have accumulated. Lift the cover or lid away from your face and arm.

- If young children want to help with meal preparation, give them something cool to mix in a location away from the cooking. Do not allow a child to stand on a chair or sit on the counter next to the stove.

- Children should not be allowed to use cooking appliances until they are tall enough to reach cooking surfaces safely. As children get older and taller and assume more cooking responsibilities, teach them safe cooking practices.

- Check all handles on appliances and cooking utensils to guarantee they are secure.

- Consider the weight of pots and pans. Attempt to move only those items that you can easily handle. Wear short sleeve or tight-fitting clothing while cooking.

- Always use oven mitts or potholders when moving pots of hot liquid or food.

- Keep pressure cookers in good repair and follow manufacturer's instructions.

- Avoid using area rugs in cooking areas, especially near the stove. If area rugs are used, ensure they have nonslip backing to prevent falls and scalds.

Scald prevention tips for the dining area

- During mealtime, place hot items in the center of the table, at least 10 inches from the table edge.

- Use nonslip placemats instead of tablecloths if toddlers are present, as young children may use the tablecloth to pull themselves up, causing hot food to spill down onto them. Tablecloths also can become tangled in crutches, walkers, or wheelchairs, causing hot liquids to spill.

Tips of careful handling of hot beverages

- Never drink or carry hot liquids while holding or carrying a child. Quick motions (reaching or grabbing) may cause the hot liquid to spill, burning the child or adult.

- Do not make hot coffee, tea, or hot chocolate in a mug that a child normally uses. Consider using mugs with tight-fitting lids, like those used for travel, when children are present. Do not place hot liquids on low coffee tables or end tables that a young child can reach.

Scald prevention tips when using microwaves

- Place microwaves at a safe height, within easy reach, for all users to avoid spills. The face of the person using the microwave should always be higher than the front of the door. All users should be tall enough to reach the microwave oven door, easily view the cooking area, and handle the food safely. Microwaves installed above counters or stoves can be a scald hazard for anyone.

- Children under age 7 should not operate the microwave unless they are closely supervised. Instruct and supervise older children.

- Never heat baby bottles of formula or milk in the microwave, especially those with plastic bottle liners. When the bottle is inverted, plastic liners can burst, pouring scalding liquids onto the baby. Always mix the formula well and test on the back of a hand or inner wrist before feeding.

- Steam, reaching temperatures greater than 200 degrees, builds rapidly in covered containers and easily can result in burns to the face, arms, and hands. Puncture plastic wrap or use vented containers to allow steam to escape while cooking. Or, wait at least 1 minute before removing the cover. When removing covers, lift the corner farthest from you and away from your face or arm.

+ Steam in microwave popcorn bags is hotter than 180 degrees. Follow package directions, allow to stand 1 minute before opening, and open bag away from the face.

+ Foods heat unevenly in microwaves. Remember, jelly and cream fillings in pastries may be extremely hot, even though outer parts feel only warm.

• Foods and liquids that have been cooked in a microwave may reach temperatures greater than boiling without the appearance of bubbling. Stir and test food thoroughly before serving or eating.

Reference

1. American Burn Association. *Scalds: A Burning Issue - A Campaign Kit for Burn Awareness Week 2000*. Online at http://www.ameriburn.org/Preven/2000Prevention/Scald2000PrevetionKit.pdf

Index